CW00547723

"This is a welcome and much-needed volume in Organization Studies, where the drive towards conformity seems relentless. Three major issues stand out: the focus on the importance of situated and plural forms of knowing in a global context; the need for empowering research by challenging mainstream epistemological commitments and values; and the need for researchers to recognise that we and those we research are human, with feelings and senses.

In a beautiful range of chapters that take us out of conventional organisational settings, the authors explore indigenous and participatory forms of knowing and researching that foreground the lived experience of participants. The chapters address important social problems from a range of methodological perspectives. For example, Devi Vijay studies the experience of palliative care through Butler's notion of vulnerability; Premilla D'Cruz, Ernesto Noronha, Saikat Chakraborty and Muneeb Ul Lateef Banday give voice to children employed as child labour through visual methods; and Srinath Jagannathan and Premalatha Packirisamy use autoethnography to reflect on their experience of becoming parents while being academics. Each chapter offers a rich narrative that stimulates you to reconsider conventional methodological values and practices".

— **Ann Cunliffe**, *Fundacao Getulio Vargas-EAESP, Brazil*

"From the title onwards you know that this is going to be an important read. This is a book that not only informs the reader of the 'how to' aspects of research methods but their emancipatory potential".

— **Albert J. Mills**, *University of Eastern Finland*

"Turning current efforts to repoliticize critical management and organisation research into an opportunity for reinvigorating the field, this brilliantly clever and thought-provoking volume is exactly what our community needs. Bell and Sengupta adeptly guide the reader through a beautiful collection of chapters, which in each their unique way speak to the kinds of researchers we want to be(come) and who we are writing for. Through honest, reflexive and responsible accounts the volume centers voices of people who are all too often othered, oppressed or exploited, bringing to the fore their lived concerns, fears or aspirations while also problematising the very concepts of 'center' and 'voice', so as not to naturalise or romanticise the speaking subject. The contributing authors not only write about the sensory, affective, embodied practice of empowering research, it is felt through the text. Individually, and in combination, these contributions offer poignant invitations to critical scholars to become more response-able, to paraphrase Emmanuel Levinas, in seeking to diversify research practices and decolonise social scientific knowledge production".

— **Sarah Louise Muhr**, *Copenhagen Business School, Denmark*

EMPOWERING METHODOLOGIES IN ORGANISATIONAL AND SOCIAL RESEARCH

This book explores the meaning and practice of empowering methodologies in organisational and social research.

In a context of global academic precarity, this volume explores why empowering research is urgently needed. It discusses the situatedness of knowing and knowledge in the context of core-periphery relations between the global North and South. The book considers the sensory, affective, embodied practice of empowering research, which involves listening, seeing, moving and feeling, to facilitate a more diverse, creative and crafty repertoire of research possibilities. The essays in this volume examine crucial themes including:

- How to decolonise management knowledge
- Using imaginative, visual and sensory methods
- Memory and space in empowering research
- Empowerment and feminist methodologies
- The role of reflexivity in empowering research

By bringing postcolonial perspectives from India, the volume aims to revitalise management and organisation studies for global readers. This book will be useful for scholars and researchers of management studies, organisational behaviour, research methodology, development studies, social sciences in general and gender studies and sociology.

Emma Bell is Professor of Organization Studies at The Open University, UK.

Sunita Singh Sengupta is Professor of Leadership and Organizational Studies at the University of Delhi, India.

EMPOWERING METHODOLOGIES IN ORGANISATIONAL AND SOCIAL RESEARCH

Edited by Emma Bell and Sunita Singh Sengupta

Routledge
Taylor & Francis Group

LONDON AND NEW YORK

First published 2022
by Routledge
2 Park Square, Milton Park, Abingdon, Oxon OX14 4RN

and by Routledge
605 Third Avenue, New York, NY 10158

Routledge is an imprint of the Taylor & Francis Group, an informa business

British Library Cataloguing-in-Publication Data
A catalogue record for this book is available from the British Library

Library of Congress Cataloging-in-Publication Data
A catalog record for this book has been requested

ISBN: 978-0-367-37058-9 (hbk)
ISBN: 978-0-367-37059-6 (pbk)
ISBN: 978-0-429-35249-2 (ebk)

DOI: 10.4324/9780429352492

Typeset in Bembo
by Apex CoVantage, LLC

CONTENTS

FIGURES

NOTES ON CONTRIBUTORS

Muneeb Ul Lateef Banday is a PhD candidate in the Organizational Behaviour Area at the Indian Institute of Management Ahmedabad, India. His research interests include discourse and subjectivity, power, politics and resistance at work, change management and critical management studies. Muneeb is currently working on his doctoral thesis on governing employee subjectivities in Indian information technology sector. He has worked on research projects examining child labour in agricultural work and change management in Indian family businesses.

Emma Bell is Professor of Organization Studies at The Open University, UK. Her research explores culture and materiality in organisations using qualitative methods of inquiry. She is also interested in how knowledge is produced and qualitative approaches to research. Recent projects have focused on the politics of management knowledge production in India and the meaning and organisation of craft work. She has authored or edited six books and published in journals including *Organization Studies, Human Relations, Academy of Management Learning & Education, Organization and British Journal of Management.*

Saikat Chakraborty is a PhD candidate in the Organizational Behaviour Area at the Indian Institute of Management Ahmedabad, India. His research interests include informal work in the formal sector, workplace dignity, labour response in precarity, organisational change and qualitative research methodologies. Saikat has seven years of work experience in human resource management, manufacturing and information technology functions in public and private sector Indian organisations before undertaking doctoral studies.

Clelia Clini is Research Associate on the project *Migrant Memory and the Postcolonial Imagination* at Loughborough University London, UK. Her research interests include migration studies; race, gender and postcolonial studies; South Asian diasporic literature and cinema. She has taught media, cultural and postcolonial studies at the American University of Rome and John Cabot University and was a postdoctoral researcher on an ESRC GCRF project on forced displacement and the arts at UCL.

Premilla D'Cruz, PhD, is Professor of Organizational Behaviour at the Indian Institute of Management Ahmedabad, India. Her research interests include workplace bullying, emotions at work, self and identity at work, organisational control and information and communication technologies (ICTs) and organisations. Premilla is the chief co-editor of the forthcoming *Handbooks of Workplace Bullying, Emotional Abuse and Harassment, Volumes 1–4*. She is the section co-editor of Labour Relations and Business Ethics at the *Journal of Business Ethics*.

Emanuela Girei, PhD, is Lecturer in Organisation Studies at the Sheffield University Management School, UK. Her research interests are in management, politics and social change. She is also interested in qualitative and critical research methodologies and processes and strategies to decolonise research and knowledge.

Jasmine Hornabrook is Research Associate on the *Migrant Memory and the Postcolonial Imagination* project at Loughborough University, UK. Her research interests include musical performance in South Asian diasporas, migration, religion and the politics of belonging in transnational music scenes. She completed her doctoral thesis at Goldsmiths, University of London, focusing on music and transnationalism within the Sri Lankan Tamil diaspora and has worked on AHRC-funded postdoctoral projects at Goldsmiths and Newcastle University.

Srinath Jagannathan teaches in the Indian Institute of Management Indore, India. His research interests are in the areas of inequality, precariousness and violence.

Emily Keightley is Professor of Media and Memory Studies at Loughborough University, UK. Her main research interest is memory, time and its mediation in everyday life. She is particularly concerned with the role of media in the relationship between individual, social and cultural memory. In 2017 she was awarded £1m Research Leadership Award by The Leverhulme Trust (2017–2022) for the project 'Migrant Memory and the Postcolonial Imagination (MMPI): British Asian Memory, Identity and Community after Partition'.

Lauren McCarthy is Senior Lecturer in Strategy & Sustainability, and Deputy Director of the Centre for Research into Sustainability at Royal Holloway University of London, UK. Lauren's research interests centre on the gender dynamics of organisational responsibilities in global value chains and the relationships between feminist social movements and corporate power. She has been published in *Business Ethics Quarterly, Business & Society, Human Relations* and *Organization Studies* amongst others.

Jennifer Manning is a lecturer in Critical Management and Critical Thinking, Strategic Management and Research Methods in Technological University Dublin, Ireland. Her research employs a decolonial feminist lens to explore colonialism, patriarchy and capitalism in management and organisation studies. Her work promotes alternative ways of working by integrating the voices, lives and experiences of those 'othered' and marginalised, particularly indigenous Global South women, into mainstream discourse. As an ethnographer, Jen has developed a critical, empowering approach to research, namely, Decolonial Feminist Ethnography.

Nita Mishra is a researcher and lecturer at the University College Cork, Ireland. Her research interests lie in rights-based approaches to development, gender and feminist methodologies. She has extensive experience working with donors and non-governmental organisations focusing on indigenous people and women in rural India. Her poetry critically acclaimed as the future of Irish feminism embodies her lived experiences working as a development researcher, and as a migrant woman, in Ireland.

Loice Natukunda, PhD, is Lecturer of HRM and Organisational Behaviour at the Department of Development studies of Makerere University, Kampala, Uganda. Her research interests are in human capital development and management in Africa applying qualitative approaches to research. She is also involved in researcher development for PhD students and early career researchers in Africa.

Ernesto Noronha, PhD, is Professor of Organizational Behaviour at the Indian Institute of Management Ahmedabad, India. His research interests include workplace dignity, labour and globalisation, workplace diversity and technology and work. He is the co-editor of the forthcoming *Asian Perspectives on Workplace Bullying and Harassment.* Ernesto is the section co-editor of Labour Relations and Business Ethics at *Journal of Business Ethics* and on the editorial committee of *Relations Industrielles/Industrial Relations.*

Premalatha Packirisamy teaches in the Tata Institute of Social Sciences, Mumbai, India. Her research interests are in the areas of gender, burnout and employee retention.

Divya Patel is a PhD candidate at the School of Management, Royal Holloway, University of London, UK. Her research explores corporate social responsibility (CSR) conceptualisation and the role of gender in executing CSR practices within co-preneurships. Her research examines CSR from feminist perspectives; most recently she has employed a feminist lens to redraw and enhance the Socio-Emotional wealth theoretical framework.

Sunita Singh Sengupta is Professor of Leadership and Organizational Studies at the Faculty of Management Studies, University of Delhi, India and former dean and head of faculty. She is the founder and Honorary Convener of the Integrating Spirituality and Organizational Leadership Foundation (ISOL Foundation), India. Her research focuses on value-based leadership and seeks to promote the development of non-violent, non-exploitative and sustainable organisations and compassionate workplace practices by drawing on the perspectives, disciplines and wisdoms enabled by diverse spiritual and religious belief systems. She was a 2007 recipient of the Homi Bhabha Fellowship and has published numerous books and articles.

Amanda Sinclair is an author, researcher and teacher in leadership, change, gender and diversity. Currently a professorial fellow, Amanda held the Foundation Chair of Management (Diversity and Change) at Melbourne Business School, 1995–2012. Her books include *Leadership for the Disillusioned* (2007); *Leading Mindfully* (2016); and, with Christine Nixon, *Women Leading* (2017). About 15 years ago, Amanda shifted gear and trained to be a yoga and meditation teacher. Much of her teaching and coaching focuses on introducing insights and practices from mindfulness to leading well. She recently completed her first fiction manuscript and – mostly – she now wants her writing to encourage people to pause, relish life, nature and the people around them.

Devi Vijay is Associate Professor at the Indian Institute of Management Calcutta (IIMC), India. Devi's research spans questions of inequality, institutions and collective action, with a specific focus on healthcare. She has co-edited *Alternative Organizations in India: Undoing Boundaries* (2018), and has published in journals including *Gender, Work & Organization*, *M@n@gement*, *Public Management Review*, *Marketing Theory* and *Journal of Marketing Management*.

1

EMPOWERING METHODOLOGIES IN ORGANISATIONAL AND SOCIAL RESEARCH

Emma Bell and Sunita Singh Sengupta

Introduction

Empowering methodologies can be understood as an ethical stance that involves creating spaces for qualitative researchers to seek to equalise power differentials 'in their relationship with research participants by paying attention to issues of voice, interpretation, interactions, dialogue, and reflexivity' (Davis 2012, 261). A key premise of empowering methodologies is that people who are othered, oppressed or exploited in or through organisations and management are invited to participate in the production of knowledge that is related to them. These people are also entitled to expect that the research process will involve some form of reciprocally beneficial exchange, for example by enabling their lived concerns, fears or aspirations to be voiced in a way which enables them to be heard, including by those who occupy positions of power.[1] In addition to systematically developing methodologies that challenge established inequalities, research empowerment relies upon an incremental practice of seeking out moments when traditional power imbalances between researchers and participants are disrupted (Ross 2017). To accomplish this, empowering methodologies draw on perspectives that engage with difference and seek to challenge oppression and inequality, including feminist (Lather 1991, 2007), critical (Alvesson and Deetz 2000), decolonial (Smith 2012) and participatory (Burns, Hyde, Killet, Poland and Gray 2014) research.

A key feature of empowering methodologies concerns their role in raising epistemological questions. For feminist methodologist Patti Lather (1991), this is founded on an examination of what it means to know, by considering 'the textual staging of knowledge' and seeking to avoid one's 'own authority from being reified' (p. 84). Empowering methodologies thus involve embracing epistemological uncertainty through the realisation that knowing is 'uncertain endeavour . . . [which involves] dealing with an uncertain world' (Morgan 1983, 386). In so doing,

DOI: 10.4324/9780429352492-1

they acknowledge that there is no such thing as value-free knowledge and allude to the inseparability of ethics and epistemology (Code 2020; Bell and Willmott 2020). Empowering methodologies thereby introduce axiological considerations related to the importance of values in producing knowledge through research. Specifically, they present a challenge to the notion of value-free science which 'simply drives values underground' (Lather 1991, 51). To enable these shifts, empowering methodologies call for reconsideration of the power relations embedded in social and organisational inquiry – for there can be no exploration of research empowerment without an understanding of power in research, including the power to silence or obscure from view.

Our commitment to empowering methodologies is prompted by concerns about the effects of colonising (Ibarra Colado 2006; Gobo 2011), Anglo-American, positivist (Üsdiken 2014; Grey 2010) and masculinised (Bell, Meriläinen, Tienari and Taylor 2020) practices of knowledge production, in the field of organisation studies. While much critical, reflexive work has been done by qualitative researchers to analyse research practice as a series of embodied, affective relationships, most mainstream research in our field adopts a positivist epistemology that assumes the existence of objective truths awaiting discovery. Values are seen as 'subjective, undermining the pursuit of truth and a potential source of bias or error' (Hiles 2012, 53). These epistemological commitments invite transactional, rational and instrumental views of research relationships which make the kinds of engagements on which empowering research relies impossible.

Our collaboration since 2011 combines two distinct and different vantage points.[2] Emma has drawn on feminist, decolonial, new materialist and qualitative methodologies and used them to rethink what it means to 'know' in organisation studies- and the purposes and consequences of such knowing (Kothiyal, Bell and Clarke 2018; Bell, Kothiyal and Willmott 2017; Bell and Willmott 2020; Bell, Winchester and Wray-Bliss 2020). Sunita has traced how concepts from Western psychology have been imported into Indian management research with a disregard for indigenous alternatives. Through her work on Indian values, spiritualities and cultural traditions (see for example Singh-Sengupta 2009, 2013), she has sought to highlight the importance of concepts of Indian spirituality – which British colonialism and Christianity attempted to destroy and appropriate – 'as sites of resistance for indigenous peoples' (Smith 2012, 78). At the same time, we seek to acknowledge the tensions that arise from our situatedness – Sunita in India and Emma in the UK. Bi-cultural (Smith 2012) research partnerships involving researchers in the global South working with those in the North are subject to power relations in a context where connections, including 'academic travel . . . patronage and sponsorship, publication and the formation of research networks . . . commonly centre on prominent figures in the metropole' (Connell 2007, 218). The publication of this book in the *Routledge India Originals*[3] series is intended as a gesture whereby we have sought to situate our collaboration, and the knowledge it has fostered, in India and the global South to a greater extent than these dynamics encourage.

The remainder of this chapter is structured as follows. We begin by explaining why we believe empowering research is urgently needed due to the current 'repositivisation' (Lather 2007) of research in a context of global academic precarity (Kothiyal et al. 2018). Next, we reflect on the situatedness of knowing and knowledge in the context of core-periphery relations between the global North and South. In the section that follows, we discuss the epistemological arguments on which empowering methodologies rely and the ethical and political purposes that they serve. We draw in particular on concepts developed by feminist philosopher and epistemologist Lorraine Code, including 'ecological thinking' (Code 2006) and 'epistemic responsibility' (Code 2020). We then consider the sensory, affective, embodied practice of empowering research which involves listening, seeing, moving and feeling, suggesting that this can facilitate a more diverse, creative and crafty (Bell and Willmott 2020) repertoire of research possibilities.[4] Finally, we identify three aspects of empowering research, showing how they relate to each of the chapters that make up this volume.

Post-positivism and repositivisation in organisational research

The assumption that knowledge can aspire to be value-free has been widely challenged by feminist, postcolonial, postmodern and critical scholars who refute the logic of scientific inquiry based on the epistemology and methodology of positivism. Some commentators argue that we are entering an era of post-positivism in the human sciences (see Prasad 2005) – a period in which the socially constituted, historically and culturally embedded and value-based nature of knowledge is recognised. Empowering methodologies are aligned with the 'methodological and epistemological ferment' that characterises 'post-positivist' human science (Lather 1991, 50). Post-positivism encourages experimentation with interactive, contextualised methods of study that are oriented towards co-constructing knowledge based on lived human experience.

Despite the ambition of the post-positivist turn, in organisation studies there has been a shift towards 'repositivisation' (Lather 2007). Thus, while there is considerable interest and diversity in qualitative research in the management disciplines, including those 'traditionally seen as founded on objectivity, "facts", numbers and quantification' (Cassell, Cunliffe and Grandy 2018, 2), the proportion of qualitative research that is published in prestigious journals continues to be low and is growing very slowly. This has been accompanied by a growing standardisation of qualitative management research where creativity has been constrained and practices have become more homogenous and formulaic (Cassell 2016). There has also been a move towards neo-positivism in qualitative organisational research. This is indicated by practices that involve demonstrating the objective validity and reliability of analytical procedures, for example statistical, inter-rater reliability checks (Cornelissen, Gajewska-De Mattos, Piekkari and Welch 2012), counting occurrences and the unreflexive use of terms like 'bias' (Bell and Thorpe 2013).

Understanding the turn towards repositivisation in organisation studies requires consideration of the contexts where knowledge is produced – the neoliberal, globalised business school. Organisational researchers face increased pressure to conform to conservative, technocratic and isomorphic norms of what counts as 'good' empirical research, often framed within a positivist or neo-positivist paradigm (Bell et al. 2017). Precarious working conditions, intensification of research and teaching and prescriptive managerial regimes mean that early career researchers are tacitly or explicitly told that critical, qualitative research is too risky, likely to be viewed as insufficiently 'systematic' and hence less likely to be published or enable academic employment (Bristow, Robinson and Ratle 2017). Practices of 'othering' qualitative research(ers) are also situated in patriarchal and colonial cultures, which position qualitative research as feminised (Mir 2018). Significant detrimental, professional and personal effects can arise for these researchers as a consequence of their failure to comply with dominant methodological norms. Ann Cunliffe (2018) offers a passionate and moving account of the oppressive effects of scientism on her identity as a qualitative organisational researcher. Her ethnographic narrative draws attention to the political and ethical consequences of her career choices over a twenty-year period in a context where 'opportunities to be imaginative and write differently are diminishing' (p. 9). Related sentiments are also expressed by researchers at the start of their research careers. Ruth Weatherall (2018) describes how she felt estranged from the normative, scientific conventions of academic writing that framed her doctoral thesis, which distanced her emotionally and ethically from women who had experienced domestic violence and had participated in her research. She urges consideration of the uneven power relationships that shape doctoral research writing and poses questions related to the kinds of researchers we want to become and who we are writing for.

The effects of uneven power relations on researcher identity are also reflected on by Rashedur Chowdhury (2017, 1111) in a discussion about his fieldwork encounters with traumatised victims and rescuers in the Rana Plaza garment factory collapse in 2013 which 'killed and injured at least 1135 and 2500 people respectively'. Chowdhury describes how his 'multiple identities' as a Bangladeshi, who had lived for 15 years in the UK, shaped 'what I think about myself, and how I am perceived by fellow academics' (p. 1113). For Chowdhury, feelings of double-consciousness resulted when 'Western society and academia . . . fails to take victims' feelings seriously'. He questions whether 'victims voices' and their 'lives, agony, and grievances' 'matter at all' in conventional research (2017, 1113–1114). Chowdhury advocates a paradigm shift in research on marginalised actors that challenges 'the narrow, orthodox way of publishing research' and instead makes 'use of oral history, literary theories, art work, and alternative philosophies' (1115–1116). As these examples powerfully attest, researchers who refuse to comply with dominant positivist norms are constituted as the 'other' who does not belong, contributing to feelings of personal and professional isolation.

Situated knowing

The contributors to this book share a concern about the need to decolonise social scientific knowledge production by translating postcolonial theory into empirical and methodological research practice. Postcolonial scholarship has done much to problematise the ontological and epistemological ground on which fields like organisation studies are based through the exposure of 'epistemic coloniality' and 'violence' (Ibarra Colado 2006; Spivak 1988). Postcolonial critiques of social scientific knowledge production trace how methods and practices (that are assumed to be universal) were, and continue to be, exported from the global North to researchers in the global South (Alatas 2003; Gobo 2011; Bell and Kothiyal 2018). In the global South, research has been a 'site of significant struggle between the interests and ways of knowing of the West and the interests and ways of resisting of the Other' (Smith 2012, 2). In considering the role of place and power in determining how knowledge is produced, Raewyn Connell (2007) traces the influence throughout the twentieth century of 'urban and cultural centres of the major imperial powers', which she refers to as the 'metropole' (p. 9), in defining the classical sociological canon. She demonstrates how sociological knowledge is linked to the imperialist gaze through the feature of 'bold abstraction' developed through the comparative method. This rests 'on one-way flow[s] of information, a capacity to examine a range of societies from the outside, and an ability to move freely from one society to another . . . features which all map the relation of colonial domination' (Connell 2007, 12). Theories are thereby claimed to be universally relevant through relating to 'social practices and human beings *in general*' (p. 34, emphasis in original). However, this assumption overlooks global inequalities in 'scientific' knowledge production that arise from European and North American imperialism. In addressing this, Connell draws attention to embodied practices of social science which she suggests may be used to challenge as well as reinforce core-periphery relations of inequality between researchers.

Linda Tuhiwai Smith suggests that even the term 'research' is inextricably linked to European imperialism and colonialism and continues, through globalisation and new forms of imperialism, to shape how knowledge is produced (Smith 2012). Smith's work traces how 'indigenous peoples', who have been constituted as 'other' as a consequence of practices of colonisation and imperialism, are objectified through research, denied a voice and subjected to the imposition of Western authority over their knowledge, languages and cultures. Starting to redress these inequalities requires key cultural ideas, beliefs and theories on which modern social science relies to be problematised, including assumptions about the nature of behaviour as causal and predictable, the hierarchical positioning of humans above nature, 'the imputing of a Western, psychological self . . . to group consciousness' (p. 77) and linear views of time.

Smith's exploration of indigenous knowledges is not simply an addition to existing ways of knowing and producing knowledge, but a means of calling into question dominant practices of knowledge production in the global North (Connell 2007).

Decolonising methodologies seek 'to ensure that research with indigenous peoples can be more respectful, ethical, sympathetic and useful' (Smith 2012, 9). This can be enabled by the use of research methods that bring to the fore and name indigenous cultural values and practices and treat them as integral to the methodology adopted. Emphasis is placed on the *process* of research which is equally, or perhaps more important than the outcome. Value is placed on '"reporting back" to people and "sharing knowledge" . . . [based on] a principle of reciprocity and feedback' (Smith 2012, 16), as well as by disseminating research to wider audiences.

Empowering research can thus be understood as a practice that relies on 'embodied alterity', through which 'we grapple with how we are both the same and different from others . . . experientially and intellectually' (Cunliffe 2018, 17), and uses this to acknowledge our ethical responsibilities to others. Crucially, by 'being open to others', including 'different ways of thinking and acting . . . without trying to integrate them into monologic forms of theorizing', it becomes possible 'to disrupt the conventional Western philosophical discourse which privileges unity and sameness over alterity and difference' (Dahl 2001, 28–29, cited in Cunliffe 2018, 18).

Epistemologies of responsibility

To understand how the diverse ambitions of empowering methodologies may begin to be realised, we turn here to the work of feminist philosopher Lorraine Code (1987/2020, 2006, 2020) and feminist methodologist Patti Lather (1991, 2007). Code (2006) argues that we need to move beyond humanistic, post-Enlightenment 'epistemologies of mastery' (p. 21), which are based on a 'narrowly conceived standard of rationality, citizenship and morality' that position and privilege '*man* at the centre of the universe' (p. 1, emphasis in original). Feminist and decolonising epistemologies are also critical of the concept of the autonomous self-determining and self-actualising self, which reflects a modernist, masculinist, Western ideal (Code 2006; Smith 2012; Alldred and Gillies 2012). Code is critical of such reductionism in promoting hegemonic conceptions of knowledge and suppressing other ways of knowing, including indigenous and traditional knowledge. Code's 'ecological subject' is 'self-critically cognizant of being part of and specifically located within a social-physical world that constrains and enables human practices, where knowing and acting always generate consequences' (Code 2006, 5).

Feminist, decolonising methodologies instead involve working with

> a materially constituted and situated subjectivity for which place, embodied locatedness, and discursive interdependence are conditions for the very possibility of knowledge and action . . . [E]cological thinking . . . naturalizes feminist epistemology's guiding question – 'whose knowledge are we talking about?'
>
> *(Code 2006, 21)*

This stance problematises the positioning of the researcher 'as "the one who knows"' (Lather 2007, 11), in a move which Lather refers to as 'getting lost':

> In postfoundational thought, as opposed to the more typical sort of mastery project, one epistemologically situates oneself as curious and unknowing. This is a methodology of "getting lost," where we think against our own continued attachments to the philosophy of presence and consciousness that undergirds humanist theories of agency.
>
> *(Lather 2007, 9)*

This radical uncertainty deeply troubles Enlightenment ideals of knowing and naming. It implies a different relationship between the researcher and the researched, characterised by reciprocity and implying 'give and take, a mutual negotiation of meaning and power' (Lather 1991, 57), as well as giving (something) back, including by enabling the voices of research participants to be heard. However, empowering methodologies must also problematise the concept of voice which can give rise to the 'romance of the speaking subject' (Lather 2007, 136).

> From the perspective of the turn to epistemological indeterminism, voice is a reinscription of some unproblematic real. This is a refusal of the sort of realism that is a reverent literalness based on assumptions of truth as an adequation of thought to its object and language as a transparent medium of reflection. The move is, rather, to endorse complexity, partial truths, and multiple subjectivities.
>
> *(p. 136)*

This enables displacement of 'the human researcher/observer from her/his central position (and hence as key arbiter) in the interaction between the world of events and the processes of research' (Fox and Alldred 2015, 1.3).

These ideas call into question the emancipatory ideal of enhancing the autonomy of those who are subject to oppression as a basis for liberating them from heteronomy. Code (2006) focuses instead on '*practices of advocacy*' as a basis for '*making knowledge possible*' (p. 165, emphasis in original). The purpose of these 'liberatory practices' is 'to (re)enfranchise epistemologically disadvantaged, marginalized, disenfranchised Others' by advocating 'in favor of the significance, cogency, validity, credibility of another person's testimony . . . [or that of] a group, institution, or society' (p. 165). The word 'testimony' is used by Code to signal the situatedness of claims to knowledge based on '*someone's* speech act' (p. 172). To be effective, advocacy relies on 'engaging in ongoing dialogue *with* the other . . . and for the other(s) . . . in places where they themselves may not be authorized, credentialed, confident enough to speak . . . or may otherwise require expert advocates' (p. 176, emphasis in original).

Code proposes ecological thinking as a way of 'remapping . . . epistemic and social-political terrains' (2006, 4) by acknowledging the interconnectedness of the human and other-than-human world in scientific and secular projects of inquiry. Ecological thinking is based on situated knowledge, as

> not just a place *from which to know* . . . indifferently available to anyone who chooses to stand there . . . [but as a] place to know whose intricacies have to be examined for how they shape both knowing subjects and the objects of knowledge.
>
> *(p. 40, emphasis in original)*

In recent work, Code returns to the notion of 'epistemic responsibility' (Code 1987), as the responsibility to know 'carefully and well' (Code 2020, 2). She asserts that this relies on consideration of '*who "we" think we are*' (Code 2020, 2, emphasis in original) and how we enact our place in the world. Exercising this responsibility requires unsettling 'lived assumptions about freedom and entitlement in the twenty-first-century Western world' (Code 2020, 2). It 'incorporates responsiveness and recognition in engaging with people, places, practices, theories, things, and situations "on their own terms," so far as this is possible' (Code 2020, 3).

By rejecting epistemic individualism, in favour of epistemic responsibility, empowering methodologies seek to interrupt taken-for-granted understandings and '"make strange" what passes for natural in epistemic practices and in the people, places, events, social arrangements, and phenomena' in ways which disrupt the status quo and 'entrenched structures of power and privilege' (Code 2006, 71). Rather than trying to secure claims by demonstrating mastery over method, and using them to produce an authoritative account that shows the researcher's control over knowledge, research is understood as messy (Law 2004), based on risky practices that must be craftily made and remade in each and every situated context of inquiry (Lather 2007; Bell and Willmott 2020).

Doing empowering research: listening, seeing, moving and feeling

Moving beyond theoretical considerations of what is meant by empowering research, we consider here how to go about doing empowering research as a sensory, affective, embodied practice that relies upon listening, seeing, moving and feeling. We show how empowering methodologies can provide a basis for 'generating and refining more interactive, contextualized methods in the search for pattern and meaning rather than prediction and control' (Lather 1991, 72).

Soyini Madison's (2018) practice of performance ethnography draws attention to the importance of listening as an ethical, affective, improvisational practice. It is through listening to one another, she suggests, that co-dependence is enabled, giving rise to the possibility of hearing 'soundings of otherness' (Fischlin 2015, 294, cited in Madison 2018). Such methods involve being in the present and paying attention with the whole body; 'it is a profound sensory experience' (p. 33) that gives rise to 'relational labour' where 'consciousness shifts from self in relation to group, to body in relation to body, to movement in relation to space and time, to past in relation to present, and to fragment in relation to developing whole' (Foster 2015, 402, cited in Madison 2018, 32). Listening thereby offers a means of 'respons(ability)' that 'unsettle[s] the expected' by responding to 'each moment [as] a potential surprise' (2018, 32). Improvisational practices play an important role in moving away from epistemologies of mastery and towards uncertainty and alterity. Listening is also crucial in moving away from practices which position the researcher as knowing and participants as known about.

However, listening is made more difficult because the performed, embodied identity of the 'successful' academic relies on demonstration of authoritative knowing in order to demonstrate proficiency as a cultural member of academic communities, including those that are labelled as critical (Bell and King 2010). The practice of listening therefore needs to be cultivated, through careful attentiveness and a spirit of openness.

Methods of narrative performance based on oral history and life story interviewing can provide a basis for transforming 'personal stories and hidden histories into public events . . . exceed[ing] verifiable statements [and becoming] living testimonies of competence [enabled by] socially transmitted and embodied techniques' (Madison 2018, 119–120). These methods make explicit the relationship between researchers as listeners and interlocuters as tellers of stories about their lives. Whether such performances are public or made for specific audiences, staged live or video recorded, scripted and rehearsed or highly improvised, involving trained actors or participants who tell their own stories, the point is that such enactments are based on 'many hours of embodied labour and ethics' (p. 122).

Sensory, affective and embodied research practices are also associated with visual, including arts-based, methods. Visual communication combines rationality with emotion through mimesis, in contrast to the dialogical qualities of written or spoken words which tell or explain, rather than show something (Bell and Davison 2013). This is especially important when working with participants who are marginalised or disadvantaged, and in dealing with sensitive topics (Kothiyal et al. 2018; McCarthy and Muthury 2016; Bell and Kothiyal 2018). Visual methods such as drawing, photography and video offer the possibility of 'meaningful participation' by providing a creative means for participants to use their voice in a way which raises awareness of, and helps to redress, power inequalities (McCarthy and Muthury 2016, 16). Photographs can also be used to invite consideration of the relations between the photographer, the photographed subject and the audience and provide a powerful means of relating to memory (Roberts 2011). Research in media and memory studies (Keightley 2017), management (Vince and Warren 2012) and sociology (Bolton, Pole and Mizen 2001) has demonstrated the value of visual methods as a way of enhancing participants' control over representations and meaning by involving them in taking photographs. Several of the authors in this book explore this potential (see chapters by Hornabrook, Clini and Keightley, Mishra, Vijay, Patel and McCarthy).

Empowering research also draws attention to bodily movement in spaces and places. For example, the walking interview method, used in health studies, anthropology and geography, is founded on the premise that what people say is shaped by the situation in which they are located. Walking through landscapes and 'chatting with participants' can prompt them into making 'connections to the surrounding environment and . . . [they may be] less likely to try and give the "right" answer' to questions that are asked (Evans and Jones 2011, 849). Similarly, 'go-along' interviewing involves researchers walking 'with interviewees as they go about their daily routines, asking them questions along the way', in a hybrid between participant observation and interviewing which is similar to the method of 'shadowing' but potentially enables greater movement between locations (Evans and Jones 2011, 850). The emphasis is on participants guiding the researcher through the 'real or virtual space' within which they conduct their life (Garcia, Eisenberg, Frerich, Lechner and Lust 2012, 1395). This can help to 'balance[e] the power dynamic inherent in research and thus encourage[e] a more collaborative approach' (p. 1396). It can be particularly useful when engaging with economically and socially disadvantaged participants, such as migrant or street workers, who lack ownership of private space. In addition, walking interview methods emphasise the sensory experience of walking through time and space (Murray and Järviluoma 2019).

Doing empowering research also relies on touch and feeling. The sense of touch, including the significance of what is experienced through the skin (Brewis and Williams 2019), and is enabled by bodily encounters, including with nonliving materials as well as other bodies, tends to be overlooked in research. The importance of touch extends beyond the haptic and is related to our capacity for emotional feeling, including the affective relations that enable us to be touched 'by what we are near' (Ahmed 2010, 30). By emphasising embodied, sensory practices of knowing which seek to transcend the limitations of language, empowering methodologies draw attention to the importance of affective encounters (Fox and Alldred 2017) in enabling greater recognition of the politics and ethics of research. As this discussion highlights, empowering research relies upon messy, creative, imaginative, improvisational practices that facilitate engagement with the elusive, multiple and continually evolving realities we encounter (Law 2004). Empowering methodologies are thereby understood as a relational engagement between participants, researchers and other living and inanimate matter through situated, embodied and affective encounters (Barad 2007; St. Pierre 2013; Fox and Alldred 2015, 2017).

However, while we have drawn attention to the importance of methods in empowering research, we simultaneously seek to distance ourselves from the assumption that 'better' methods enable 'better', or more 'real', accounts of lived experience (Lather 2007). The use of any method must be tempered by a concern to avoid 'methodology-as-technique', where the practice of the method becomes fetishised and mastery or 'methodolatry' (Chamberlain 2000), as a demonstration of scientific rigor, becomes more important than the empirical story that is told (Bell et al. 2017). Empowering methodologies thus build on a commitment to gathering and respecting empirical evidence 'while recognizing that evidence cannot speak for itself, but achieves its status *as* evidence' (Code 2006, 47, emphasis in original), or does not, depending on the social and power relations within which evidence is embedded.

Three aspects of empowering research

In this concluding section of the chapter, we outline three aspects of empowering methodologies that connect to the chapters in this volume. The first concerns the importance of empowering methodologies as an umbrella concept (Girei and Natukunda, this volume) which can be 'used loosely to encompass and account for a set of diverse phenomena' (Hirsch and Levin 1999, 200). This is similar to a 'boundary object' (Fox 2011, 71–73) which has the power to 'speak' to different audiences. By having 'the capacity to be understood by actors in more than one setting' empowering methodologies enable establishment of a 'shared language' that can be used by actors in different knowledge communities to challenge dominant, hegemonic research practices. The concept of empowering methodologies also enhances translation of research 'across culturally defined boundaries' and 'between communities of knowledge or practice', including practitioner and scholarly communities. In so doing, and like other related umbrella concepts such as 'research-as-craft' (Bell and Willmott 2020), empowering methodologies can be used to drive methodological pluralism and to focus attention on ethics and politics.

A second aspect of empowering methodologies explored in this book concerns the importance of reflexivity (Girei and Natukunda, Jagannathan and Packirisamy, Mitra, Sinclair, all this volume). As Morgan (1980, 373) argues, 'when we engage in research action, thought and interpretation, we are not simply involved in instrumental processes geared to the acquisition of knowledge but in processes through which we actually make and remake ourselves as human beings'. When researchers enter into fieldwork relationships they, as well as research participants, are exposed to the possibility of self-reflection and identity work, including as a consequence of forming new relationships or having one's views, beliefs and ways of knowing called into question (Coffey 1999; Bell and Taylor 2014). Reflexivity offers 'a way of foregrounding our moral and ethical responsibility for people and for the world around us' (Cunliffe 2016, 741). This brings responsibilities to treat others as irreplaceable, rather than as a means to an end. It requires that we understand knowledge as intersubjectively constructed by appreciating that we are never separate from others, and who and how we are is always experienced relationally, through our encounters with others (Cunliffe 2016).

A final aspect of empowering methodologies that characterises the contributions to this volume involves moving beyond method, as stylised, routinised and accepted ways of doing research (Abbott 2004). Preoccupation with method encourages the fetishisation of technique, or methodolatry (Bell et al. 2017), whereby the demonstration of recognised methods of data collection and analysis, according to predetermined quality standards, becomes more important than the process of doing research and the relational encounters this enables. 'Methodology' provides a more encompassing term (Mir 2018; Wolcott 1991). It acknowledges the 'inextricably and unavoidably historically, politically, and contextually bound' (Fontana and Frey 2005, 695) nature of research practices which are embedded in research communities that position them normatively as 'the right way to do things' (Abbott 2004, 13). As several contributors to this volume argue (Manning, Vijay), empowering research is characterised by 'embodied locatedness', as an alternative to the 'stringent dictates of an exaggerated ideal of scientific knowledge making' (Code 2006, 20). We now briefly summarise each of the chapters in turn.

Girei and Natukunda discuss their experience of doing research in sub-Sahara Africa while navigating dilemmas linked to the legacy of colonisation and wider asymmetries that characterise South/North relations. By reflecting on identity, power and knowledge production and exploring how these dynamics shape the production of management knowledge in Africa, they propose two strategies that seek to decolonise management knowledge and research. First, a move towards more open-ended approaches to doing research, and second, a commitment towards transforming participants' role in the research process. These are intended as a means of shifting the positioning of researchers away from being seen as sources of data and towards becoming full partners in knowledge production and consumption.

Manning proposes decolonial feminist ethnography as an empowering, ethically engaged methodology that can address the complexities of the lived world and the complications of power in research to bring forward different worldviews, knowledges and lived experiences. She describes how she integrated decolonial feminist theory into a critical ethnography of marginalised, indigenous Maya women who work in backstrap weaving groups in rural, remote Highland communities in Guatemala. She argues that this practice can help to achieve epistemological decolonisation by enabling research that challenges inequality, power and politics, and recognises the intersections of voice, place and privilege. Manning asserts that this relies on dialogic, embodied performance through which power is shared and knowledge is produced together.

The chapter by Vijay reflects on the possibility of vulnerability as praxis in studies of social suffering. Drawing on research examining the community-based palliative care movement in Kerala, she discusses what it means to focus on vulnerability as an ethico–political imperative in the research process. Specifically, she explores how from a condition of vulnerability, it is possible to adopt three modes of praxis: (1) *vulnerability as susceptibility*, which allows openness to silence and challenges epistemic certitude; (2) *vulnerability as collective care*, which acknowledges the role of time and generosities; and (3) *vulnerability as learning to be affected by difference*, where one learns from the wounding and the unsettled habitations that arise over the course of fieldwork.

D'Cruz, Noronha, Chakraborty and Banday detail an empirical study that used drawings to study the lived experiences of bullied children from the Bhil community who work on cottonseed farms in India. They describe their use of the 'draw followed by talk' technique to study the organisational phenomenon of workplace bullying. The authors' account reveals the utility of drawings in enabling children to share their lifeworld, which would otherwise have been inaccessible to them. Through this participative research experience, the children felt at ease in sharing their experiences of work. The technique enabled them to give voice to the children and place them at the centre of their research, rather than treating them as passive respondents. The chapter also highlights the practical as well as ethical issues involved in using drawings in research that involves children.

In the following chapter, Hornabrook, Clini and Keightley observe that researching memories of painful pasts poses particular methodological challenges. Their study of the articulation of memories of the 1947 Partition of British India, and associated processes of migration, therefore required a participant-centred approach. Creative methods have the potential to address difficulties in elicitation through participation and collaboration. The chapter presents a creative methodological approach that was developed in the Migrant Memory and Post-colonial Imagination project. They explore how cooking, sewing and photography can evoke memories, create opportunities for participants to approach feelings of belonging, discrimination and marginalisation in oblique ways, and provide safe spaces for their articulation. The authors argue that creative methods provide spaces for collaborative and nuanced understandings of the role memory plays in community and belonging in South Asian diasporic communities.

The focus on visual methods developed in the previous two chapters is further developed by Patel and McCarthy, who discuss their use of visual participatory research methods. Reflecting on a case study into corporate social responsibility (CSR) in Gujarat, India, they show how women and men's experiences of CSR, their meaning-making and relationships within patriarchal, caste-based systems can be surfaced through freehand, participant-led drawing. They suggest that drawing, alongside verbal reflection, can play an important role in beginning the challenging, yet ethically imperative process of empowering research. Yet these processes are complicated by our own standpoints as researchers, as cultural 'insiders' and 'outsiders', and by 'being' women in the data generation process.

A highly reflexive turn is taken in the final three chapters of the book, starting with Jagannathan and Packirisamy, who present their autoethnographic narratives of becoming a mother and a father while working as academics. By focusing on personal experiences as an epistemic resource, they outline strategies for using autoethnographic sensibility as a methodological framework. Through their autoethnographic accounts of mother/fatherhood and academic work, they show how this research strategy can yield important personal insights connected to ethical knowledge and confronting the ideological normalisation of inequality.

Mishra's chapter analyses her experiences as an ethnographic researcher studying the lived experiences of women workers employed in a government food entitlement programme in India. The ethnographic fieldwork took place in Odisha State, India, and involved communities in 64 villages. Drawing upon feminist methodologies, she reflects on the fluid relationships between herself as a researcher and the women participants and explores how researcher-participant relationships are continuously shaped and re-shaped by the field site, and the sites within herself. Vijay argues that honesty, humaneness and humility in the field are more important than collecting data. She contends that qualitative researchers should keep a flexible approach and allow participants to influence the researcher's position in ways which make the research process dynamic and empowering for both.

Finally, Sinclair charts her journey as a researcher, particularly her work exploring women and leadership. Through personal stories and reflections, she maps her efforts to firstly do, then write and, finally, be, in research, differently. She shows how she felt excluded, and then wanted to differentiate herself from, what she has increasingly come to see as gendered, often oppressive, understandings of the templates, processes and purposes of 'good' research. Sinclair's teachers in learning to do and write research differently have been research participants, colleagues, students and readers, as well as feminist and critical writers, each challenging her to rethink who and how she wants to be as a researcher.

The distinctiveness of the chapters in this book arises from the connections made by the contributors, between epistemology, ontology and axiology, and methodologies and methods in doing empowering research. Taken together, they expand the limits of possibility in organisational and social scientific research and enable a move away from epistemologies of mastery (Code 2006), and towards an epistemology of uncertainty (Snitow 2015; Bell and Taylor 2014) and alterity (Cunliffe 2016). This is enabled by discussion of the significance of empowering research in contexts of their own research practice, including by providing self-reflexive accounts of research projects and collaborations. Crucially, the emphasis is on exploring the messy nature of empowering research and acknowledging that it is likely to be impossible to achieve complete empowerment, whilst emphasising the political and ethical importance of seeking to do so.

Notes

1 It could be argued that the term 'emancipation' might be preferable to 'empowerment' as a way of describing management research that seeks to challenge cultural and linguistic conventions to which people are subordinated through the 'acts of powerful agents . . . for the benefit of certain sectional interests at the expense of others' (Alvesson and Deetz 2000, 9). However, while empowering research is also concerned with identifying and analysing the 'causes of powerlessness . . . [and] systemic oppressive forces', Lather argues that critical intellectuals who focus on emancipation are in danger of assuming a 'vanguard role' characterised by 'crusading rhetoric' through being 'stuck in a framework . . . [that] sees the "other" as the problem for which they are the solution' (Lather 1992, 131–132). As this critique highlights, 'espousing an emancipatory model of research has not of itself freed researchers from exercising intellectual arrogance or employing evangelical and paternalistic practices' (Smith 2012, 180). While the concept of empowerment is also open to this possibility, we use it here to signal a need to strike a balance between *a priori* theory that informs the value of emancipation against the need to be 'respectful of, the experiences of people in their daily lives' and resist the 'dangers of imposing researcher definitions on [an] inquiry' (Lather 1991, 54, 78).

2 This was enabled by funding from UK-India Education Research Initiative (UKIERI) Exchange 'Building international capacity in management research' and Research Project 'Developing Empowering Methodologies in Management Research', UGC 17–18–10.
3 See www.routledge.com/collections/7648 (Accessed 10.04.19).
4 Similar calls have been made in sociology, where it has been argued that there is a need for 'live methods' which are more 'artful and crafty' (Back and Puwar 2012, 9).

References

Abbott, A. (2004) *Methods of Discovery: Heuristics for the Social Sciences*. New York: Norton & Co.

Ahmed, S. (2010) Happy Objects. In M. Gregg and G.J. Seigworth (Eds.), *The Affect Theory Reader*. London: Duke University Press, pp. 29–51.

Alatas, S.F. (2003) Academic Dependency and the Global Division of Labour in the Social Sciences. *Current Sociology*, 51: 599–633.

Alldred, P. and Gillies, V. (2012) Eliciting Research Accounts: Re/producing Modern Subjects? In T. Miller, M. Birch, M. Mauthner and J. Jessop (Eds.), *Ethics in Qualitative Research*. 2nd ed. London: Sage.

Alvesson, M. and Deetz, C. (2000) *Doing Critical Management Research*. London: Sage.

Back, L. and Puwar, N. (2012) A Manifesto for Live Methods: Provocations and Capacities. In L. Back and N. Puwar (Eds.), *Live Methods*. Oxford: Wiley/Blackwell: The Sociological Review, pp. 6–17.

Barad, K. (2007) *Meeting the Universe Halfway: Quantum Physics and the Entanglement of Matter and Meaning*. London: Duke University Press.

Bell, E. and Davison, J. (2013) Visual Management Studies: Empirical and Theoretical Approaches. *International Journal of Management Reviews*, 15(2): 167–184.

Bell, E. and King, D. (2010) The Elephant in the Room: Critical Management Studies Conferences as a Site of Body Pedagogics. *Management Learning*, 41(4): 429–442.

Bell, E. and Kothiyal, N. (2018) Ethics Creep from the Core to the Periphery. In C. Cassell, A. Cunliffe and G. Grandy (Eds.), *SAGE Handbook of Qualitative Business and Management Research Methods*, London: Sage, pp. 546–561.

Bell, E., Kothiyal, N. and Willmott, H. (2017) Methodology-as-technique and the Meaning of Rigor in Globalized Management Research. *British Journal of Management*, 28(3): 534–550.

Bell, E., Meriläinen, S., Tienari, J. and Taylor, S. (2020) Dangerous Knowledge? The Political, Personal and Epistemological Promise of Feminist Research in Management and Organization Studies. *International Journal of Management Reviews*, 22(2): 177–192.

Bell, E. and Taylor, S. (2014) Uncertainty in the Study of Belief: The Risks and Benefits of Methodological Agnosticism. *International Journal of Social Research Methodology*, 17(5): 543–557.

Bell, E. and Thorpe, R. (2013) *A Very Short, Fairly Interesting and Reasonably Cheap Book about Management Research*. London: Sage.

Bell, E. and Willmott, H. (2020) Ethics, Politics and Embodied Imagination in Crafting Scientific Knowledge. *Human Relations*, 73(10): 1366–1387.

Bell, E., Winchester, N. and Wray-Bliss, E. (2020) Enchantment in Business Ethics Research. *Journal of Business Ethics*. https://doi.org/10.1007/s10551-020-04592-4

Bolton, A., Pole, C. and Mizen, P. (2001) Picture This: Researching Child Workers. *Sociology*, 35(2): 501–518.

Brewis, D. and Williams, E. (2019) Writing as Skin: Negotiating the Body in(to) Learning about the Managed Self. *Management Learning*, 50(1): 87–99.

Bristow, A., Robinson, S. and Ratle, O. (2017) Being an Early Career CMS Academic in the Context of Insecurity and 'Excellence': The Dialectics of Resistance and Compliance. *Organization Studies*, 38: 1185–1207.

Burns, D., Hyde, P., Killet, A., Poland, F. and Gray, R. (2014) Participatory Organizational Research: Examining Voice in the Co-production of Knowledge. *British Journal of Management*, 25: 133–144.

Cassell, C. (2016) European Qualitative Research: A Celebration of Diversity and a Cautionary Tale. *European Management Journal*, 34(5): 453–456.

Cassell, C., Cunliffe, A. and Grandy, G. (2018) Introduction: Qualitative Research in Business and Management. In C. Cassell, A. Cunliffe and G. Grandy (Eds.), *Sage Handbook of Qualitative Business and Management Research Methods*. London: Sage, pp. 1–32.

Chamberlain, K. (2000) Methodolatry and Qualitative Health Research. *Journal of Health Psychology*, 5: 285–296.

Chowdhury, R. (2017) Rana Plaza Fieldwork and Academic Anxiety: Some Reflections. *Journal of Management Studies*, 54(7): 1111–1117.

Code, L. (1987/2020) *Epistemic Responsibility*. Albany: SUNY Press.

Code, L. (2006) *Ecological Thinking: The Politics of Epistemic Location*. Oxford: Oxford University Press.

Code, L. (2020) *Manufactured Uncertainty: Implications for Climate Change Skepticism*. Albany: SUNY Press.

Coffey, A. (1999) *The Ethnographic Self: Fieldwork and the Representation of Identity*. London: Sage.

Connell, R. (2007) *Southern Theory*. Cambridge: Polity Press.

Cornelissen, J., Gajewska-De Mattos, H., Piekkari, R. and Welch, C. (2012) Writing up as a Legitimacy Seeking Process: Alternative Publishing Recipes for Qualitative Research. In G. Symons and C. Cassell (Eds.), *Qualitative Organizational Research*. London: Sage, pp. 184–203.

Cunliffe, A.L. (2016) 'On Becoming a Critically Reflexive Practitioner' Redux: What Does It Mean to *Be* Reflexive? *Journal of Management Education*, 40(6): 740–746.

Cunliffe, A.L. (2018) Alterity: The Passion, Politics and Ethics of Self and Scholarship. *Management Learning*, 49(1): 8–22.

Dahl, D.E. (2001) *Originary Passivity: Selfhood and Alterity in Ricoeur and Levinas*. Unpublished Master's Thesis. University of Guelph, Canada.

Davis, C. (2012) Empowerment. In L.M. Given (Ed.), *The Sage Encyclopedia of Qualitative Research Methods*. London: Sage, p. 261.

Evans, J. and Jones, P. (2011) The Walking Interview: Methodology, Mobility and Place. *Applied Geography*, 31: 849–858.

Fischlin, D. (2015) Improvised Responsibility: Opening Statements (Call and) Responsibility: Improvisation, Ethics, Co-creation. In R. Caines and A. Heble (Eds.), *The Improvisation Studies Reader: Spontaneous Acts*. London: Routledge, pp. 289–295.

Fontana, A. and Frey, J.H. (2005) The Interview: From Neutral Stance to Political Involvement. In N. Denzin and Y.S. Lincoln (Eds.), *The Sage Handbook of Qualitatitive Research*. 3rd ed. Thousand Oaks: Sage.

Foster, S.L. (2015) Improvisation in Dance and Mind. In R. Caines and A. Heble (Eds.), *The Improvisation Studies Reader: Spontaneous Acts*. London: Routledge, pp. 398–403.

Fox, N.J. (2011) Boundary Objects, Social Meanings and the Success of New Technologies. *Sociology*, 45(1): 70–85.

Fox, N.J. and Alldred, P. (2015) Inside the Research-assemblage: New Materialism and the Micropolitics of Social Inquiry. *Sociological Research Online*, 20(2): 6. https://doi.org/10.5153/sro.3578. www.socresonline.org.uk/20/2/6.html

Fox, N.J. and Alldred, P. (2017) *Sociology and the New Materialism: Theory, Research, Action*. London: Sage.

Garcia, C.M., Eisenberg, M.E., Frerich, E.A., Lechner, K.E. and Lust, K. (2012) Conducting Go-along Interviews to Understand Context and Promote Health. *Qualitative Health Research*, 22(1): 1395–1403.

Gobo, G. (2011) Glocalizing Methodology? The Encounter Between Local Methodologies. *International Journal of Social Research Methodologies*, 14: 417–437.

Grey, C. (2010) Organizing Studies: Publications, Politics and Polemic. *Organization Studies*, 31: 677–694.

Hiles, D.M. (2012) Axiology. In L.M. Given (Ed.), *The Sage Encyclopaedia of Qualitative Research Methods*. London: Sage, pp. 53–56.

Hirsch, P.M. and Levin, D.Z. (1999) Umbrella Advocates Versus Validity Police: A Lifecycle Model. *Organization Science*, 10: 199–212.

Ibarra Colado, E. (2006) Organization Studies and Epistemic Coloniality in Latin America: Thinking Otherness from the Margins. *Organization*, 13: 463–488.

Keightley, E. (2017) *Memory and the Management of Change: Repossessing the Past*. London: Macmillan.

Kothiyal, N., Bell, E. and Clarke, C. (2018) Moving Beyond Mimicry: Creating Hybrid Spaces in Indian Business Schools. *Academy of Management Learning & Education*, 17(2): 137–154.

Lather, P. (1991) *Getting Smart: Feminist Research and Pedagogy With/in the Postmodern*. New York: Routledge.

Lather, P. (1992) Post-critical Pedagogies: A Critical Reading. In C. Luke and J. Gore (Eds.), *Feminism and Radical Pedagogy: A Reader*. London: Routledge, pp. 120–137.

Lather, P. (2007) *Getting Lost: Feminist Efforts Towards a Double(d) Science*. Albany: State University of New York.

Law, J. (2004) *After Method: Mess in Social Science Research*. London: Routledge.

Madison, D.S. (2018) *Performed Ethnography and Communication: Improvisation and Embodied Experience*. Abingdon: Routledge.

McCarthy, L. and Muthury, J.N. (2016) Engaging Fringe Stakeholders in Business and Society Research: Applying Visual Participatory Research Methods. *Business & Society*: 1–43.

Mir, R. (2018) Embracing Qualitative Research: An Act of Strategic Essentialism. *Qualitative Research in Organizations and Management*, 13(4): 306–314.

Morgan, G. (1983) Part III: Conclusions. In G. Morgan (Ed.), *Beyond Method: Strategies for Social Science*. Newbury Park: Sage, pp. 368–376.

Morgan, G. and Smircich, L. (1980) The Case for Qualitative Research. *Academy of Management Review*, 5(4): 491–500.

Murray, L. and Järviluoma, H. (2019) Walking as Transgenerational Methodology. *Qualitative Research*: 1–10.

Prasad, P. (2005) *Crafting Qualitative Research: Working in the Post-Positivist Traditions*. Armonk, NY: M.E. Sharpe.

Roberts, B. (2011) Photographic Portraits: Narrative and Memory. *Forum: Qualitative Social Research Sozialforschung FQS*, 12(2): Art. 6. www.qualitative-research.net/index.php/fqs/article/view/1680/3203

Ross, K. (2017) Making Empowering Choices: How Methodology Matters for Empowering Research Participants. *Forum: Qualitative Social Research Sozialforschung FQS*, 18(3): Art. 12. http://doi.org/10.17169/fqs-18.3.2791

Singh-Sengupta, S. (2009) *Developing Intra Psychic Health: Insights from Indian Tantra System*. Unpublished Homi Bhabha Research Report.

Singh-Sengupta, S. (2013) The Ontology and Epistemology of Spirituality in Organizational Studies: A Concept Note Based on Hindu Philosophy. *International Journal on Vedic Foundations of Management*, 1(2): 150–154.

Smith, L. T. (2012) *Decolonizing Methodologies: Research and Indigenous Peoples*. 2nd ed. London: Zed Books.

Snitow, A. (2015) *The Feminism of Uncertainty: A Gender Diary*. Durham, NC: Duke University Press.

Spivak, G. (1988) Can the Subaltern Speak? In C. Nelson and L. Grossberg (Eds.), *Marxism and the Interpretation of Culture*. Urbana: University of Illinois Press, pp. 271–313.

St. Pierre, E.A. (2013) Post Qualitative Research: The Critique and the Coming After. In N.K. Denzin and Y.S. Lincoln (Eds.), *Collecting and Interpreting Qualitative Materials*. Los Angeles, CA: Sage Publications, pp. 611–625.

Üsdiken, B. (2014) Centres and Peripheries: Research Styles and Publication Patterns in 'Top' US Journals and Their European Alternatives, 1960–2010. *Journal of Management Studies*, 51: 764–789.

Vince, R. and Warren, S. (2012) Participatory Visual Methods. In C. Cassell and G. Symon (Eds.), *Qualitative Organizational Research: Core Methods and Current Challenges*. London: Sage, pp. 275–295.

Weatherall, R. (2018) Writing the Doctoral Thesis Differently. *Management Learning*, 50(1): 100–113.

Wolcott, H. (1991) *Ethnography: A Way of Seeing*. Walnut Creek: AltaMira Press.

2

DECOLONISING MANAGEMENT KNOWLEDGE AND RESEARCH

Reflections on knowledge, processes and actors

Emanuela Girei and Loice Natukunda[1]

Introduction

In this chapter, drawing on our experience as management researchers in East Africa, we reflect on key dilemmas, questions and principles that have shaped our commitment to decolonise how we plan and do research. Acknowledging the variety of meanings and practices associated with this commitment, here we broadly understand it as a way of doing research that is underpinned by the commitment to identify, expose and challenge the ways in which colonialism and imperialism continue to shape knowledge production. We thus understand decolonising knowledge as a process of "disengagement from the colonial syndrome" (Loomba 1998, 21) through which we address the legacy of colonialism and decolonisation in both the coloniser and the colonised, in the metropolis and the colony. Such a view is relevant for three reasons. Firstly, for its emphasis on the temporal dimension: the words "syndrome" and "process" allude to a lasting yet dynamic condition and invite us to see decolonisation as a continuous tension and effort throughout the research journey. Second, Loomba's view calls into question both the coloniser and the colonised, thus calling for shared responsibility and engagement, a process with people, which involves the 'researcher' as much as the 'researched'. Third, the term 'disengagement' creates space for personal and collective agency, in the sense that while recognising the exercise of power in knowledge production and its detrimental effects on non-Western systems of knowledge and identities, it opens up possibilities for re-configuring such an oppressive dynamic.

This chapter emerges from our reflections on how we have engaged in this process and how this engagement has been shaped by our subjective positions, some of them overlapping, other distant and many others in between these two poles. We met in 2016 in the UK, in a university management school where Loice had just finished her PhD and Emanuela had just started working as a lecturer and

DOI: 10.4324/9780429352492-2

soon started talking about our experiences of doing management research (and prolonged empirical work) in East Africa. Loice was born and has lived a significant part of her life in Uganda, interspersed with long periods spent in Europe, especially for study and work purposes. Emanuela was born in Southern Europe and has been studying and living in the UK for several years and has spent a prolonged period in East Africa for study and work purposes.

Since meeting, we have continued sharing our reflections around identities, power and knowledge production, to deepen our understanding of their various impacts on our research. This chapter emerges from this journey, and while our main empirical experience is in East Africa, here we shall expand our reflections so as to consider more widely how divides between the Global South and the Global North impact upon management research and what we can do to avoid reproducing and possibly altering such a divide through our research.

This chapter starts by acknowledging that methodological dilemmas related to conducting management research in Africa can be daunting for researchers who come from other continents, and also for scholars based in Africa (see for example Lunn 2014). Among the various reasons for this, two are particularly important for this chapter.

Firstly, and as discussed in the introduction, dominant paradigms commonly indicate that research is/should be objective with an unseen researcher and hidden "subjects" and thus data needs to be quantified (Keane, Khupe and Seehawer 2017). As far as conducting research in Africa is concerned, this objectivist, positivist approach is considered more scientific in contrast to attempts to explore phenomena in specific settings using qualitative approaches that directly engage with participants and context. This is reflected in the predominantly quantitative and comparative approach that has characterised the study of Africa (Cheeseman, Death and Whitfeld 2017). It is thus not surprising that management and organisational knowledge production in Africa have been dominated by quantitative studies with researchers settling for surveys with the support of managers and research assistants for data collection. In addition, the African continent is often considered a "field" from which to mine "raw data" (Nhemachena, Mlambo and Kaundjua 2016, 15). These features are not unique to management and organisation studies and can be better understood as aspects of the predominantly extractive nature of research on/in the Global South.

Secondly, and linked to the previous point, several scholars have emphasised that management and organisation studies is predominantly produced by and for the Global North (Murphy and Zhu 2012), largely neglecting the African continent. They have questioned not only the assumed universality of management knowledge but especially its relevance to and for African organisations (Girei 2017). More generally, knowledge generated about the Global South is often published in peer-reviewed journals which are largely inaccessible not only to the communities and organisations involved in the research but also to local policymakers and academics (Melber 2017)

In this chapter, starting from these epistemological, ethical and political dilemmas and with the aim of exploring linkages, overlaps and controversies of research practices committed to both decolonial and emancipatory principles, we address three main questions: What is decolonised knowledge? How can it be produced/generated/constructed? Who can produce/generate/construct it? Importantly, we do not attempt the impossible task of offering exhaustive answers to these questions, but rather, we understand them as heuristic strategies that guide our reflections on decolonising management knowledge and research.

In the following sections, we address these three questions in turn before developing some concluding reflections. As may be noticed, the three questions elicit reflections that intersect, producing overlaps and common themes. It is, for example, difficult to discuss the question of how knowledge is decolonised without mentioning who produces/generates/constructs it.

Unpacking 'decolonised knowledge'

In addressing the question "What is decolonised knowledge?", we engage with some existing notions including 'indigenous knowledge', 'embodied locatedness' and 'radical contextualisation'. But firstly, we reflect on some of the key tenets that nurture theoretical debates associated with decolonising knowledge and research. These key tenets have in common a preoccupation with knowledge and power, or more precisely with exploring, enlightening and subverting how knowledge and power are intimately connected. Many of these critiques refer directly or indirectly to Foucault's suggestion that truth, rather than being an intrinsic property of certain ideas and notions, depends on, and is determined by, the exercise of power. More precisely, Foucault argues that "the exercise of power perpetually creates knowledge and, conversely, knowledge constantly induces effects of power. It is not possible for power to be exercised without knowledge, it is impossible for knowledge not to endanger power" (1980, 52). Thus, Foucault argues, power and knowledge are mutually constitutive and together shape the "regime of truth" of a certain historical period and/or society, thus sanctioning which discourses are true, who has the authority to speak about them and which are truth's validation criteria (Foucault 2009). In this sense, knowledge is neither innocent nor neutral but rather a product of specific historical and political contexts and conditions. This suggests that notions of objectivity, universality and neutrality, which are still considered crucial features of scientific production and instrumentally invoked to elevate it above other knowledge systems, should be considered, in themselves locally constructed. Rather than being intrinsic attributes of scientific knowledge, they have become dominant as a result of European expansion (Harding 1997). Furthermore, to achieve such self-declared superior status, Western knowledge systems have systematically silenced dissident voices and suppressed alternative knowledge systems. But such censorship has been "neither accidental nor benign" (Collins 2000, 3); indeed, it has been instrumental in creating the myth of Western superiority and maintaining the unequal power relations, such as those between the West and the rest of the world. Instrumental censorship is particularly evident if we look at dominant narratives about Africa and/or African management, as highlighted in the following.

Academic knowledge about Africa, whatever the field, is mainly produced by institutions and scholars located outside the continent (Mama 2007). Secondly, there are disciplines in which Africa is virtually absent, and management was for a long time one of these (Murphy and Zhu 2012). More generally, it can be argued that management research and management knowledge (and thus management curricula) are largely based on Western epistemologies, ontologies and empirical cases (see also Wiegratz 2009), which continue to reinforce and reproduce the epistemic violence that underpins them.

Significantly, even the sub-field of cross-cultural management seems to ignore the entire continent of Africa (Fougère and Moulettes 2011), and on the rare occasions when it is mentioned, it is often done with embarrassing superficiality (Nkomo 2011). This does not mean that academic production in Africa is scarce. Rather, it means that African scholars are under-represented and/or marginalised in/by the main global knowledge production centres, such as top-tier journals or internationally renowned conferences, and that rich indigenous academic production remains largely ignored in Northern academic circles. Indeed, around 2012, key authors on the subject argued that a notable increase in publications on management as a subject was undeniable although they admitted that high-impact publications published far less about management and organisational contexts in Africa than in the West and Asia (Kamoche, Chizema, Mallahi and Kahindi 2012, 1).

The silencing of African academic perspectives (and more generally of the Global South) is also reproduced through other institutionalised mechanisms, such as the peer-review process, namely the process that determines what gets published, or, in Foucauldian terms, what knowledge is and who has the power to legitimise it. For instance, a recent report (Publons 2018) provides a global picture of the scientific reviewing process, drawing on data from Publons, Web of Science and ScholarOne manuscripts databases. The report distinguishes between established and emerging regions, with the former (11 in total) including the US, some European countries and Australia and Japan, while the latter (9 in total) includes China, Brazil and India. While there are sharp asymmetries within both regions (as for instance in the case of the US dominating editors and reviewers' statistics within the established group), if we make inter-region comparisons, it emerges that, across all academic fields, 96% of the editors come from established regions and they tend to disproportionally choose reviewers from their own regions. More generally, the report highlights that emerging regions are significantly under-represented in the review process, thus suggesting that wittingly or unwittingly the same process that should legitimise and uphold good scientific standards is fully complicit with the historically rooted silencing of perspective coming from outside the dominating Western academic circles.

Another way in which instrumental censorship has been reproduced is through misrepresentation of former colonies' histories, citizens and identities and ways of knowing and working. In this respect, and more generally with regard to the relationship between power and knowledge, particularly illuminating is Said's 'Orientalism' (Said 2003) and Mudimbe's work 'The Invention of Africa' (1988).

Drawing on Foucault's understandings of knowledge, power and truth, Said's and Mudimbe's works unveil how knowledge about the Other served to justify and legitimise Western expansion and domination. For instance, the "colonial library" (Mudimbe 1988, 175) – the degrading representation of Africans in anthropologists', missionaries' and explorers' texts – systematically and instrumentally portrayed Africans as incapable of being in charge of their own lives. Mudimbe (1988) argues that this repertoire provided the ideological explanation and legitimisation for forcing Africans into colonialism concealed under the guise of a salvation/civilising mission. Such distortions are neither unique to anthropology nor of the past; instead, they can also be found in contemporary management knowledge.

For instance, if we look at management and organisation studies, previous research suggests (Prasad 2009) that if a non-Western organisation does not deploy practices and/or policies that are considered "normal" in the West, it is considered deficient, lacking and in need of modernisation or innovation. Similarly, if a non-Western organisation employs practices or policies unknown in the West, they are considered traditional or "ethnic" and usually, again, in need of innovation.

The problem here is what could be called the "epistemic colonialism" (Ibarra-Colado 2006) of Western-centric perspectives, which continue to approach organisations located outside the West through a comparative lens, emphasising how they differ from or resemble the Western model. This, in turn, nurtures and reinforces historically rooted assumptions regarding the superiority and universality of Western management perspectives and organisational practices.

Towards decolonised knowledge

Starting from the acknowledgement of issues around knowledge, power and colonialism, several scholars have attempted to envisage what decolonised knowledge might be. More recently there have been calls for a turn towards so-called "indigenous" knowledge (see for instance Darley and Luethge 2019). However, starting from the assumption that "all knowledge is local knowledge" (Okere, Njoku and Devisch 2005), our views resonate with critiques that suggest the notion of indigenous knowledge lacks clear definition and is often enclosed in a romanticised aura, based more on anecdotes than rigorous research (Jackson 2013). Often the notion of indigenous/local knowledge is founded on a neglect of how global and local dynamics and structures of oppression and domination shape local knowledge production. Dominant understandings of indigenous knowledge often privilege cultural dimensions, neglecting material and structural elements and dynamics. As Mohan and Wilson suggest (2005) idealisations of local knowledge usually do not consider how both neoliberalism and state power contribute to perpetuating marginalisation and impoverishment and, as such, unduly overstate the role that local knowledge can play for/in emancipatory goals.

Secondly, local communities (across the world) are not homogeneous but are rather stratified according to several dimensions, such as for instance gender, education, marital status and age. Such stratifications not only influence and shape what can be said and thought but especially by whom. For instance, Emanuela worked for several months with a small NGO in Eastern Africa, which was highly internally stratified and where the presence/absence of the director significantly determined what could be said and who could speak. A member of staff clearly told her: "In this organisation you can only agree with the director". In this sense, what might be seen as consensual, shared, locally "true" might mask embedded and interiorised social controls (Kothari 2001).

Furthermore, the notion of indigenous knowledge is used to contrast the local with the global, but this is highly problematic because it neglects intersections between cultures and ways of knowing. For instance, it has been argued that Africa and the West are much more intermingled and internally diversified than is assumed by such binary logics as local/global or Africa/the West (Appiah 2007); their ways of knowing are both tainted by their encounter, thus neither of them can claim to be completely pure (Quayson 1997). Acknowledging the intricateness of Africa and the West opens spaces for empowerment[2] as it implies rejecting narratives of tout-court domination and calls for more attention to the ways through which the colonised resisted, subverted and re-articulated dominant narratives. Other scholars have also argued that this inward focus on indigenous knowledge, often advocated by Southern scholars, might further aggravate the South/North divide and the isolation experienced by many scholars in the Global South, with the risk of "enclaves becoming not a refuge and bastion for independent thought, but isolated and intellectually starved islands" (Hamann et al. 2020, 6).

In this sense, and in line with the notion of "embodied locatedness" (Code 2006) discussed in the introductory chapter of this book, we find it more productive to employ the notion of "radical contextuality" (Escobar 2008, 200), which calls for a stronger embodiment and embedment of context(s) (human, cultural, symbolic, historic, economic and so on). Along similar lines, other scholars emphasise the need to establish a radical research approach that encourages researchers to deeply engage with participants and local communities and to draw on their contexts instead of relying on Western perspectives and standards (Ruggunan 2016; Dyll 2020; Keeyaa 2020). For example, if one looks at the literature on organisational development (OD) in Africa, several studies examine whether its methodology and assumptions can be transferred and applied to African contexts (Golembiewski 1991). The main problematic aspect of these studies, we suggest, is that while committed to scrutinise universalisation of Western management knowledge, they continue to use it as an authoritative lens through which to approach organisations around the world. For instance, some studies (see James 2004; Lewis 2002; Jackson 2003), acknowledging OD's North American origin, seem driven by questions such as whether OD fits within African organisations and culture, thus using OD, a Western artefact, as the analytical lens through which to investigate African contexts. A radically different approach for those engaged in OD projects would be that of understanding what are the meanings associated with this concept (if any) and be willing to co-develop approaches and contents that are meaningful and relevant to the context where the organisation works.

Such an engagement with local contexts would contribute to a threefold benefit. Firstly, this would result in a widening of epistemological and ontological perspectives, insofar as it would encourage greater mingling among perspectives and knowing systems that, as mentioned, according to some scholars have always occurred.

However, what is also needed is to make this process visible and to expose the ways in which what is called "Western" is often the result of centuries of oppressive and exploitative relations between the Global North and the Global South (Cooke 2003). This process implies engaging with alternatives ways of producing knowledge and meanings, and making visible hybridised conceptualisations and practices. For example, even though the bureaucratic procedures of managing organisations in Africa were introduced from a different context, they were found to be informed by informal traditional values in Loices's case study at a development organisation in Uganda. The supervisors seemed to collude with employees in subverting the formal bureaucratic rules but then return to such formal guidelines when problems arose or a decision was disputed. There are usually what Argyris and Schon (1974) called espoused theories or what people value and what they actually do (or theories-in-use). Exposing these processes in management research might help enlightening how organisations and workers in the Global South actively engage with and re-articulate Western managerial doxa on the basis of their cultural and political values and material conditions.

A second benefit resulting from a stronger engagement with local contexts would be superseding the comparative lens that often underpins research about the Global South. For instance, as Mamdani highlights, whatever the specific field, a key methodological quandary concerns "the legitimacy of Africa as a unit of analysis" (Mamdani 2004, 8) and the need to abandon the comparative lens that has been consistently adopted to approach Africa, where Africa is conceptualised only according to the ways it differs from the West/ern canon. In this sense, radical contextuality implies a design and an analytical lens focused on, and responsive to, locally grounded meanings and understandings. We understand this as a continuous process that accompanies all research, but especially empirical and interpretative studies that are based on self-reflective practice, an issue which we shall return to later in the chapter.

A final important outcome of the researcher's stronger engagement with local contexts is the potential for a greater local impact so that research is not merely written for other researchers within an academic club (Keane et al. 2017). Several scholars, especially those committed to participatory and/or activist methodologies, go even further, suggesting not only that should participants be key beneficiaries of the research but also that asymmetrical researcher-researched relationships should be abandoned in favour of a research setting where different actors act as co-researchers, as we shall explore further in the next section.

How do we produce decolonised knowledge?

In this section we reflect on what we can practically do to decolonise our research approach, and we shall focus on two strategies that, on the basis of our empirical experiences and reflection on them, we think can be particularly helpful. One of them refers to how we conceptualise the overall and overarching research design, the other to how we conceptualise participants in our projects.

Liberating the research design

A key principle that underpins scientific research as it is usually taught in research methods modules at all levels regards the importance of a deliberately constructed research design, which tends to assume that the researcher is able to plan and direct what will happen in the different phases of study, thus obfuscating the role that indeterminacy and contingency play in social research processes. This linear, determinist understanding of the research process is enforced globally through institutional practices, such as those put in place to disburse research funding and to grant ethical clearance (De-Tardo Bora 2004). These widely accepted practices usually require the researcher to be able to identify clear aims, questions and methods before the research starts and to inform participants in advance about their contributions and potential risks. We use these examples to highlight that generally accepted "gold standards" of good scientific research not only adversely impact upon qualitative, inductive research in general Cunliffe (2010), they also have specific detrimental consequences for research that is committed to decolonising knowledge. As discussed in the previous section, a critical feature of decolonised research is engagement with local contexts and meanings, local epistemologies and ways of knowing. However, we suggest that these engagements are profoundly compromised when the research design, its questions, methodologies and expected impact must be determined before the research starts, as is often expected by grant application or ethical clearance procedures. In other words, we suggest that such rigorous planning adversely impacts the possibility of continuing to nurture intermingling of ways of knowing, engagement with local contexts and silenced and/ or neglected epistemologies. For instance, Emanuela worked as advisor for many years with organisations in Europe and her approach had been shaped for a long time by principles and practices of action research and action learning. When she started working with NGOs in Eastern Africa she thought that those principles could still constructively guide her work, but it came soon apparent that even the simple principle of planning regular collective cycles of action a reflection sometimes clashed with local practices and dynamics. Sometimes, the team-level was not the best setting for reflecting and learning, in other occasions the context was highly unpredictable (in rural areas, heavy rain was enough to disrupt agreed schedules and plans), and more generally the line between adhering to a methodology, albeit inspired to participation and equality, and prescriptive expectations became increasingly blurred. Indeed, there are some accounts of research projects where the researcher transparently acknowledges the constraining impact that of structured research design. For instance, Marshall and McLean, reflecting on one of their experiences of co-operative inquiry, clearly acknowledge that "they (we) were asking [the participants] to play their (our) game" and that throughout the project they took an increasingly "central and expert role" (Marshall and McLean 1988,

219–220). Others have exposed divergent understandings and needs of researchers and participants in relation to participation (Busza 2004; Arieli et al. 2009). These studies highlight how a structured research design might constrain and hinder the engagement with research practices, epistemologies and knowledge systems alternative to those prescribed/required by Western mainstream research practices.

Acknowledging the constraints that might arise by adhering to a structured research design, we suggest that efforts to decolonising knowledge and research would benefit from a stronger orientation towards an open-ended approach. This means freeing the research from rigid design, theoretical perspectives or methods, so as to minimise the risk of imposing categories and of constraining thinking and actions. An open-ended approach is underpinned by a disenchanted attitude towards faithfulness to a single grand theory and/or a single overarching paradigm, with a correspondingly increased commitment to perspectives that are locally and historically situated (Denzin and Lincoln 2003). Drawing on transformational studies, Alff and Hornidge (2019) suggest that an open-ended orientation contributes to the decolonisation of knowledge and research by emphasising the "'becoming' rather than the state of being" (ibid,148) of social formations and actors, as well as the interconnections and overlapping among social actors, disciplines and spaces. Such an understanding of open-ended orientation resonates with what in the introduction to this volume is defined as research-as-craft (Bell and Willmott 2020); it invites us to appreciate the emergent dimension of the research process and its inherent indeterminacy. We believe that an open-ended orientation concerning both empirical investigation and theoretical framing helps in widening perspectives, learning from the persons involved in the research and avoiding, as much as possible, constraining the variety and unpredictability embedded in human actions and social phenomena.

Liberating research roles

It can be argued that in Western scientific practices, participants are often reduced to mere objects of knowledge for those with the power to do research (Colignon 2004) and extract useful data. Looking specifically at Africa, it has been argued that the continent is commonly seen as a "field" from which to mine "raw data" (Nhemachena et al. 2016, 15), with little attention for local views, perspectives or even research impact. Wright (2002) suggests that a reconfiguration of relations between researchers and the researched, so to ease participants' stronger engagement in the research process, not only has the potential to empower indigenous researchers and communities but would enable the production of knowledge that is meaningful and usable in and for the context where the research took place. As Smith (1999) puts it:

> When indigenous people become the researchers and not merely the researched, the activity of research is transformed. Questions are framed differently, priorities are ranked differently, problems are defined differently, and people participate on different terms.
>
> *(Smith 1999, 193)*

In this sense, we suggest considering an orientation towards co-production as both an epistemological and ethical stance that contributes to subverting the pattern of silencing and misrepresentation of southern voices and perspectives, but also to assert the right of human beings to shape the knowledge about them, their communities and their organisations (Reason and Bradbury 2001, 2).

However, participatory methodologies are multi-layered, ontologically dynamic and can have manipulatory intents, and as such, they should not be seen as necessary emancipatory. For example, participation could be seen opportunistically as a means to achieve a goal already set, where thus participation is used to legitimise decisions outside participants' sphere of influence (Taylor 2001).

Another important controversy surrounding participatory methodology, especially when employed in the global South, concerns its associations with a romanticised idea of local communities or indigenous organisations, which, by ignoring power relations, can result in a strengthening of pre-existing local structures of domination (Hildyard, Hegde, Wolvekamp and Reddy 2001). In this sense, the adoption of participatory methodologies needs to be understood in close connection with radical contextuality and its emphasis on strong engagement with local contexts, paying attention to the various positionalities and dynamics of their social, economic and political fabrics. We understand such an engagement with local contexts as a process that requires consideration of the timeframe available for specific research and reflection on to what extent it allows such a deep engagement with local contexts. As social scientists and management researchers, we are all aware of the intricateness and indeterminacy of social settings, and this can constructively nurture our reflections on the value and limitations of the engagement we are able to enact in our research projects.

A meaningful way to practice decolonisation and deepen our understanding of local contexts is to develop stronger and long-term partnerships with local researchers (Khupe and Keane 2017; Keikelame 2018; Datta 2018). This is as long as we shy away from the "cultural translator" role's expectations and avoid reproducing the asymmetries which often shape South/North research partnerships (Parker and Kingori 2016; Carbonnier and Kontinen 2015), where high-income countries continue to be the prime beneficiaries of international research collaborations, especially with regard to authorship, international profile and influence in the research process. It has been noted that North-South research collaborations are often understood primarily as a tool to ease data generation. Parker and Kingori (2016) use the insightful expression of "glorified field worker" (ibid,4) to depict the role of researchers from low-income countries, suggesting that their role in global research projects is often relegated to the empirical phase, without a meaningful involvement in the other phases of the research project. Turner (2010) goes further speaking of "ghost researchers", whose role is rendered invisible and whose work is appropriated by colleagues in the Global North and hardly ever recognised in authorships and intellectual property rights. Despite the dearth of literature on the lived experience of research assistants and collaborators in law-income countries (Turner 2010), a recent study by Sukarieh and Tannock (2019) enlightens the alienation and self-estrangement often felt by researchers in the Global South working in international partnerships, which arise not only from the lack of meaningful involvement in the research process but also from the fact that research agendas are usually alien to their interests or those of the communities and organisations they research. This not only deprives researchers from the Global South the opportunity to make more substantial and meaningful contributions but also undermines the potential that such collaborations embed to question the Western canon and contribute to decolonise knowledge and research practices.

There are a number of relevant publications (see for instance Matenga, Zulu, Corbin and Mweemba 2021; Erdal, Amjad, Bodla and Rubab 2015) and websites (see for instance Rethink Research Collaborative https://rethinkingresearchcollaborative.com/) that provide useful suggestions on how to make research partnerships more equitable, and common across them is the attention towards fair distribution of resources, transparency and accountability and commitment towards shared agendas and goals. However, it is also important to acknowledge that some of the constraints to equal North/South research collaborations stem from historical geopolitical asymmetries, material conditions and the working of the research industry. In our own experience of collaboration, sometimes we can only navigate through but not transform existing divides. A simple, but significant, example in this sense concerns access to academic resources, such as books and journals, which is often profoundly unequal in South/North collaborations. Similarly, on several occasions, Emanuela, with the financial support of her institution, presented joint work to international conferences whose fees were often inaccessible for Loice (and the great majority of researchers in the Global South). More broadly, if we look at wages, distribution of resources and power over research agendas, it becomes apparent that there is the need to advocate for fundamental changes in how global academic research is regulated and produced. This reminds us the limits of short-term interpersonal engagements among researchers; we certainly have a role to play in making research partnerships more equal also by contributing to sustainable transformative changes within academic and research institutions in the Global North. In this sense, Erdal et al. (2015) invite us to look at inequalities in international research not as a scandal, which would call for denunciation and clamour, but as a "skandalon" (an ancient Greek word that means stumbling block), which requires constant analysis and reflections and consequent conscious choices regarding our practices and commitments, at micro and macro levels, at home and abroad, with our international partners and with our colleagues and managers in our own institution.

Engagement and self-reflexive practices

Our understanding of researcher engagement in the research process is not only intellectual but also emotional, ethical and political. Epistemologically, this choice is based on the perspective that rejects the objectification of people, recognising the intrinsically social and political nature of knowledge. As Heron puts it, researchers "can't get outside, or try to get outside, the human condition in order to study it" (1996, 20), so their efforts should be directed to grounding their accounts in their own experience. Feminist scholars have also stressed the epistemic value of the full engagement of the researcher in the production of knowledge and the importance of interconnecting emotions, ethics and reasoning. Collins (2000) points out that her training as a social scientist was not adequate to study the knowledge of subjugated people, such as that of Black African women, as they have developed distinctive and alternative ways of generating and validating knowledge which are missing from and ignored by dominant knowledge validation structures and institutions (2000, 252).

Decolonising research requires a rupture with requirements for "suprapolitical objectivity" (Said 2003, 10), on the basis of the acknowledgement that knowledge cannot be neutral, being necessarily interwoven with ethical and political threads. In other words, if it is not possible to detach knowledge from the circumstances where it is produced, from the necessary involvement of the researcher in a society, it follows that it is crucial for researchers to continuously investigate whom and what specific research choices and actions they are serving. Furthermore, researchers should reflect on which truths and worldviews they sustain and their impact on the lives of those involved in the research, as well as the impact of their identities and positionalities in the processes of knowledge production.

Thus, our understanding of research committed to decolonisation begins by acknowledging the partial, subjective and local value of the narrative texts we generate. Our accounts are inevitably constrained by the "subjective positions available" to us (Holvino 1996, 530) and are inevitably generated through and embedded in our standpoint and positionalities, i.e., in our embodied identities and in those positioned by others (Franks 2002), both constantly negotiated and re-defined throughout the research process. Our standpoints and positionalities are themselves "objects" of the research process, through which we gain a deeper understanding of our roles and knowing strategies.

In this sense, self-reflexive practice for us entails a critical analysis of the assumptions, values and interests underpinning and guiding our action and thinking which implies not only self-awareness about them but also availability to question them (Alvesson and Deetz 2000, 112–113). This critical analysis does not aim to reach a neutral perspective, disengaged and detached from our own subjectivity; rather it signals an interest and commitment to minimise the manipulative intents and effects of pre-given lens (see for instance Wray-Bliss 2003) and to strengthen opportunities for a genuine engagement with the diversity of voices and perspectives we encounter. In this sense, a kind of self-reflexive practice that we have found particularly helpful in our practices is inspired by Foucault's notion of problematisation (Deacon 2000). Here we use the term to indicate an attitude curious and sceptical of too familiar interpretations and thinking, which invites us to critically investigate what we take for granted and why what we consider relevant and why and what we dismiss and why. These reflections are part of self-reflective practice, which entails scrutinising the working of our subjectivity along the research process. More precisely, starting from the acknowledgement of our active role along the whole research process, from the construction of the empirical material to the production of the final text, self-reflexive practice helps to blur the divide between researcher and researched inasmuch as it constantly requires to include ourselves in our investigation.

Self-reflexive practice entails continuous attentiveness towards our identities and those (we perceive) of research participants and other people we encounter, paying attention to what and how we disclose things about ourselves, how we position each other and especially how all these factors are intertwined in the process of knowing. Yet, our understanding of self-reflexivity blurs also the inner/outer dichotomy, for, drawing on Coffey (1999), it invites us to reflect not only on how my stances, assumptions and identities have influenced the whole research but also on how they have been negotiated and reshaped because of and along the research process, depending on the contexts, the encounters made and the events that occur during the research process.

Producing decolonised knowledge

We start from the acknowledgement that any commitment to decolonise research cannot eschew exploring how our identities shape the research process. Among various angles that can be employed in such a journey, one particularly relevant for this research refers to the insider/outsider dimension, which is widely recognised as a shaper of research processes (Girei and Natukunda 2020). We look at how being an insider or outsider can shape the knowledge generation process and the impact on researchers' commitment to decolonise knowledge.

Insider/outsider identities

In exploring issues around researchers' identity when doing research in former colonies, it is enlightening to consider academic debates on this matter in those countries and continents. For instance, within African scholarship issues of identity and alterity/Otherness/Africanness have been at the centre of African intellectual production since the beginning of the 1900s (Mudimbe 1988; Appiah 1992), and significant sections of these debates concern the impact and role of African identity in shaping the production of knowledge about Africa.

Some influential perspectives, developed within wider intellectual and political debates such as Negritude, Black Consciousness, Pan-Africanism have constructed/ celebrated an African collective identity clearly distinguished from, and in opposition to, Western/white identity, and as such, better suited to understand and interpret local contexts and develop knowledge that is responsive and meaningful to those contexts. Acknowledging that these perspectives have been critically scrutinised for their perpetuation of colonial binary thinking and essentialistic-racist categories (Appiah 1992; Mbembe 2002), two reflections within these debates are particularly enlightening in relation to identity and the decolonisation of knowledge.

As Mudimbe suggests, the intellectual debate around Africanness of the 1950s–1960s mainly represents the search for a new epistemic foundation of the discourse on Africa, which puts at its centre history and African historicity. In this sense, the contribution of Negritude or Pan-Africanism is not so much about the notion of African collective identity; instead, it resides in their endeavour to "postulate a major anthropological stance, nobody is at the centre of human experience" (Mudimbe 1988, 175–200). In this sense, Negritude's and Pan-Africanism's focus on authenticity and tradition exposes the will to rescue history, systematically denied in Western narratives, where Africa "was a mere object within European historiographies" (Mudimbe 1988, 177). This reflects what we discussed previously regarding the centrality of "radical contextuality" (Escobar 2008) in developing an orientation that is careful to avoid and tries to supersede the comparative lens that has characterised research in Africa and more generally in former colonies.

A caveat that we need to consider with regard to Negritude and Pan-Africanism is that they were (and are) foremost anti-colonial and anti-imperial movements (Shivji 2006) that are committed to tackling oppression and injustice and establishing the right of self-determination. In this sense, the focus on racial distinctiveness was an "anti-racism racism [toward] the abolition of racial difference" (Sartre 1965, 18), thus what today can be called "strategic essentialism", i.e. "a strategic use of positivist essentialism in a scrupulously visible political interest" (Spivak 2006, 281). Indeed, and especially since the 1950s and throughout the 1960s and 1970s, what Zeleza calls the "golden age" of African universities (Zeleza 2009, 112), intellectual and political movements overlapped almost completely, inasmuch it would be virtually impossible to detach the intellectual debate on Africanness from the struggle for liberation from colonial domination first, and for auto-determination after. In this sense, decolonising research cannot eschew a reconfiguration of roles and asymmetries, starting from the divide between the researcher-researched and the hierarchies that shape North-South research collaborations. Such a reconfiguration would enable research to be sensitive and responsive to principles of liberation and self-determination, thus learning from, and contributing to, advancing key reflections on identity and knowledge.

Problematically, in management and organisation studies, including mainstream cross-cultural management, insider and outsider identities have been chiefly understood on the basis of the nationality and/or national culture. In this regard, Mahadevan (2011) suggests that other cultures that co-constitute our identities (e.g. those linked for instance to our work, family, political and/or religious faith to name a few) might be as equally powerful as the assumedly most important national culture. Thus, identities can go beyond nationality and geographical classification to cover a range of identities including age, caste, ethnicity, religious belief, sexuality, physical ability, personality and even class (Tinker and Armstrong 2008, 53).

Furthermore, our identities and affiliations are also shaped and positioned, by those around us (including research participants and colleagues), and we certainly cannot assume any clear correspondence between how our identities are perceived by ourselves and the other. Thus, rather than assuming a fixed and stable national identity, we find more productive to engage with issues of insiderness and out-siderness conceiving them as social and situational and continuously intersecting with several axes of identity, such as, for instance, class, gender, skin colour and ethnicity. Here we are, however, not proposing or supporting an additive or mul-tiplicative understanding of intersectionality, according to which different axes of identities can each be conceptualised, operationalised and studied separately (Cor-lett and Mavin 2014). As Liu points out, intersectionality has often been used (and especially in management studies) as "a tool for collating and commodifying 'dif-ferences'" (2018, 88). This neglects to explore interlocked systems and dynamics of power and oppression and as such can undermine efforts to address racism and marginalisation, while simultaneously reinforcing white privileges and domination (Rodriguez and Freeman 2016).

Thus, while we suggest that intersectionality might help supersede an under-standing of identity and culture as fixed and stable and thus a rigid divide between insider and outsider researchers, we think it is crucial to accompany these reflec-tions with acknowledgement of the relevance of issues of race and racism, whiteness and blackness and wider asymmetries. Such asymmetries have historically framed the relations between indigenous African populations and outsiders, and between African scholars and outsider scholars in academic knowledge production. In this sense, while current debates have contributed to developing an understanding of (African) identity as "multiple, fluid, historically and institutionally constructed along various dimensions of difference" (Mama 2007, 15) others have emphasised the relevance (if not necessity), of endogenous and independent African knowl-edge (Mkandawire 1999; Adesina 2008).

Thus, while we embrace the epistemological and political value of intersection-ality, we are also cautious and attentive not to obscure historically rooted and still existing asymmetries, including those shaped by race and skin colour. Our view is that rather than exploring whether, and how being an insider or outsider ease the production of knowledge, it would be more productive to self-reflect on how these statuses are constructed and contested in the research process and how they intersect with other identity's axes.

It is worth noticing that despite the relevance of issues of race and skin colour, it is still very rare to find management scholars' reflections/contributions on the impact of these dimensions of their own research practice. Thus, while in this section we highlight the dangers of rigid conceptualisation of insiders and outsiders, we think it is important to recognise different roles for research-ers that come from abroad and researchers that live where the research takes place but also the histories they come from and with. With this regard, our own experiences and reflections suggest that in research committed towards emanci-patory change, insider and outsider researchers might, and maybe should, take different roles insofar as being organically linked with the research's participants widens and strengthens possibilities for a long-term commitment, which is usu-ally a necessary precondition for research to be able to contribute to sustainable change.

Conclusion

In this chapter, guided by three questions, we have shared our reflections on our engagement with decolonising knowledge and research, looking specifically at Africa, but, we believe, addressing some of the key issues related to knowledge production and validation in the Global South and Global North. As specified in the introduction, we employed three overlapping questions as a heuristic strategy to explore reflections and debates that have so far nurtured our commitment to decolonise our research. We have, in particular, argued for the need to abandon traditional understandings of scientific research, in favour of research approaches that have a more open-ended orientation, focused on the situated, emergent and constructed nature of knowledge. Our invitation to embrace "radical contextuality" should, however, not be easily translated into a search for local, indigenous knowledge. Rather, we think that a more productive way to understand the engagement with local contexts is based not only on engaging with different meanings and understandings, wary of the indissolubility among power and knowledge, at all levels, including the often romanticised "local". We, however, think that an in-depth entanglement with the contexts where the research takes place is a crucial condition for decolonising knowledge. This implies not only an engagement with new and diverse meanings, ways of knowing and working but also a reconfiguration of prescribed roles and accepted dynamics between insider and outsider researchers and participants. Finally, in our practice, we have also found engaging with debates and reflections on our identities and how these shape the research process and our role particularly enlightening. In this chapter we have aimed at sharing our reflections on our journey towards decolonised knowledge and research practices, which is in itself a journey with no ending, or, as Datta suggests, an "on-going process of becoming, unlearning, and relearning regarding who we are" (Datta 2018, 2).

Notes

1 Alphabetical order, both authors contributed to the chapter equally.
2 Acknowledging the contestations and controversies around the notion of empowerment, in line with the book's premises (see the introduction), we employ the term with two important caveats. Firstly, we understand 'empowerment' as a boundary-object, namely an artefact that, despite, and because of, its various meanings, is able to glue different narratives and audiences, beyond existing controversies, thus creating space for concrete actions (Courpasson, Dany and Clegg 2011, 815). Secondly, we see empowerment as a liberatory process and goal through which legitimise and endorse knowledge actors, epistemologies and ontologies silenced by the Western scientific canon.

References

Adesina, J. O. (2008) Archie Mafeje and the pursuit of endogeny: Against alterity and extroversion. *Africa Development* XXXIII(4): 133–152.
Alff, H. and Hornidge, A. (2019) 'Transformation' in international development studies: Across disciplines, knowledge hierarchies and oceanic spaces. In I. Baud, E. Basile, T. Kontinen and S. Von Itter (Eds.), *Building Development Studies for the New Millennium.* Cham: Palgrave Macmillan.

Alvesson, M. and Deetz, S. A. (2000) *Doing Critical Management Research*. London: Sage.

Appiah, K. A. (1992). *In My Father's House. Africa in the Philosophy of Culture*. Oxford: Oxford University Press.

Appiah, K. A. (2007) *The Ethics of Identity*. Princeton, NJ: Princeton University Press.

Argyris, C. and Schon, D. (1974) *Theory in Practice: Increasing Profesional Effectiveness*. San Francisco: Jossey-Bass Publishers.

Arieli, D., Friedman, V. J., et al. (2009) The paradox of participation in action research. *Action Research* 7(3): 263–290.

Bell, E. and Willmott, H. (2020) Ethics, politics and embodied imagination in crafting scientific knowledge. *Human Relations* 73(10): 1366–1387.

Busza, J. (2004) Participatory research in constrained settings: Sharing challenges from Cambodia. *Action Research* 2(2): 191–208.

Carbonnier, G. and Kontinen, T. (2015) Institutional learning in north-south research partnerships. *Revue Tiers Monde* 1(1): 149. https://doi.org/10.3917/rtm.221.0149

Cheeseman, N., Death, C. and Whitfeld, L. (2017) Notes on researching Africa. *African Affairs, Virtual Issue on Research Notes*. Available online only, 1–5. https://doi.org/10.1093/afraf/adx005.

Code, L. (2006) *Ecological Thinking: The Politics of Epistemic Location*. Oxford: Oxford University Press.

Coffey, A. (1999) *The Ethnographic Self: Fieldwork and the Representations of Identity*. London: Sage Publications.

Colignon, A. (2004) Postcolonial theory and organizational analysis: A critical engagement by anshuman prasad review by: Richard A. *The Academy of Management Review* 29(4): 702–704.

Collins, P. (2000) *Black Feminist Thought: Knowledge, Consciousness and the Politics of Empowerment*. New York: Routledge.

Cooke, B. (2003) The denial of slavery in Management Studies. *Journal of Management Studies* 40(8): 1895–1918.

Corlett, S. and Mavin, S. (2014) Intersectionality and identity: Shared tenets and future research agendas for gender and identity studies. *Gender in Management: An International Journal* 29(5): 258–276. ISSN 1754–2413.

Courpasson, D., Dany, F. and Clegg, S. (2011) Generating Productive Resistance in the Workplace. *Organization Science* 23(3), pp. 801–819.

Cunliffe, A. L. (2010) Crafting qualitative research: Morgan and Smircich 30 years on. *Organisational Research Methods*: 1–27.

Darley, W. K. and Luethge, D. J. (2019) Management and business education in Africa: A post-colonial perspective of international accreditation. *AMLE* 18: 99–111.

Datta, R. (2018) Decolonizing both researcher and research and its effectiveness in indigenous research. *Research Ethics* 14(2): 1–24.

Deacon, R. (2000) Theory as practice: Foucault's concept of problematization. *Telos* 2000(118): 127–142.

De-Tardo Bora, K. A. (2004) Action research in a world of positivist-oriented review boards. *Action Research* 2(3): 237–253.

Denzin, N. K. and Lincoln, Y. S. (2003) Introduction: The discipline and practice of qualitative research. In N. K. Denzin and Y. S. Lincoln (Eds.), *The Landscape of Qualitative Research. Theories and Issues*. Thousand Oaks, CA: Sage, pp. 1–45.

Dyll, L. (2020) Methods of "literacy" in indigenising research education: Transformative methods used in the Kalahari. *Critical Arts* 33(4–5): 122–142. https://doi.org/10.1080/02560046.2019.1704810

Erdal, M. B., Amjad, A., Bodla, O. Z. and Rubab, A. (2015) *Equality in North-South Research Collaboration*. https://blogs.prio.org/2015/07/equality-in-north-south-research-collaboration/. Accessed 10th October 2020.

Escobar, A. (2008) *Afterword. The New Development Management. Critiquing the Dual Modernization. S. Dar and B. Cook., 198–203*. London: Zed Books.

Foucault, M. (1980) *Power/Knowledge: Selected Interviews and Other Writings, 1972–1977*, Ed. C. Gordon. New York: Harvester Wheatsheaf.

Foucault, M. (2009) *The Archeology of Knowledge*. Oxon: Routledge.

Fougère, M. and Moulettes, A. (2011) Disclaimers, dichotomies and disappearances in international business textbooks: A postcolonial deconstruction. *Management Learning* 43: 5–24.

Franks, M. (2002) Feminisms and cross-ideological feminist social research: Standpoint, situatedness and positionality – Developing cross-ideological feminist research. *Journal of International Women's Studies* 3(2): 1–13.

Girei, E. (2017) Decolonising management knowledge: A reflexive journey as practitioner and researcher in Uganda. *Management Learning* 48(4): 453–470.

Girei, E. and Natukunda, L. (2020) Exploring outsider/insider dynamics and intersectionalities: Perspectives and reflections from management researchers in sub-Saharan Africa. In J. Mahadevan, H. Primecz and L. Romani (Eds.), *Critical Cross-Cultural Management: An Intersectional Approach to Culture*. London: Routledge.

Golembiewski, R. T. (1991) Organizational development in the Third World: Values, closeness of fit and culture-boundedness. *The International Journal of Human Resources Management* 2(1): 39–53.

Hamann, R., Luiz, J., Ramaboa, K., Khan, F., Dhlamini, X. and Nilsson, W. (2020) Neither Colony Nor Enclave: Calling for dialogical contextualism in management and organization studies. *Organization Theory*, January 2020.

Harding, S. (1997) Is modern science an ethnoscience? Rethinking epistemological assumptions. In C. E. Eze (Ed.), *Postcolonial African Philosophy: A Critical Reader*. Cambridge, MA: Blackwell Publishers, pp. 45–70.

Heron, J. (1996) *Co-operative Inquiry: Research into the Human Condition*. London: Sage.

Hildyard, N., Hegde, P., Wolvekamp, P. and Reddy, S. (2001) Pluralism, participation and power: Joint Forest Management in India. In B. Cooke and U. Kothari (Eds.), *Participation, the New Tyranny?* London: Zed Books.

Holvino, E. (1996) Reading organization development from the margins: Outsider within. *Organization* 3(4): 520–533.

Ibarra-Colado, E. (2006) Organization studies and epistemic coloniality in Latin America: Thinking otherness from the margins. *Organization* 13(4): 463–488.

Jackson, T. (2003) *Cross-Cultural Management and Ngo Capacity Building. Why Is a Cross-Cultural Approach Necessary?* INTRAC, Praxis Note No.1: 1–10.

Jackson, T. (2013) Reconstructing the indigenous in African management research. *Management International Review* 53(1): 13–38.

James, R. (2004) Exploring Od in Africa: A response to David Lewis. *Nonprofit Management & Leadership* 14(3).

Kamoche, K., Chizema, A., Mallahi, K. and Kahindi, N. A. (2012) New directions in the management of human resources in Africa. *The International Journal of Human Resource Management* 23(14): 2825–2834.

Keane, M., Khupe, C. and Seehawer, M. (2017) Decolonising methodology: Who benefits from indigenous knowledge research. *Educational Research for Social Change (ERSC)* 6(1): 12–24.

Keeyaa, C. (2020) Decolonising ethics frameworks for research in Africa. *Africa at LSE* (08 Jan 2020), pp. 1–6. Blog Entry.

Keikelame, M. J. (2018) 'The tortoise under the couch': An African woman's reflections on negotiating insider-outsider positionalities and issues of serendipity on conducting a qualitative research project in Cape Town, South Africa. *International Journal of Social Research Methodology* 21(2): 19–230.

Khupe, C. and Keane, M. (2017) Towards an African education research methodology: Decolonising new knowledge. *Educational Research for Social Change* 6(1): 25–37.

Kothari, U. (2001) Power, knowledge and social control in participatory development. In B. Cooke (Ed.), *Participation: The New Tyranny?* London: Zed Books.

Lewis, D. (2002) Organization and management in the third sector. Toward a cross-cultural research agenda. *Nonprofit Management & Leadership* 13(1): 67–83.

Liu, H. (2018) Re-radicalising intersectionality in organisation studies. *Ephemera* 18(1): 81–101.

Loomba, A. (1998) *Colonialism/Postcolonialism*. New York: Routledge.

Lunn, J. (2014) *Field Work in the Global South: Ethical Challenges and Dilemmas*. London: Routledge Taylor & Francis.

Mahadevan, J. (2011) Engineering culture(s) cross sites: Implications for cross-cultural management of emic meanings. In H. Primecz, L. Romani and S. Sackmann (Eds.), *Cross-Cultural Management in Practice*. Cheltenham: Edward Elgar Publishing Limited, pp. 89–100. https://doi.org/10.4337/9780857938725

Mama, A. (2007) Is it ethical to study Africa? Preliminary thoughts on scholarship and freedom. *African Studies Review* 50: 1–26.

Mamdani, M. (2004) *Citizen and Subject: Contemporary Africa and the Legacy of Late Colonialism*. Kampala: Fountain Publishers.

Marshall, J. and McLean, A. (1988) Reflection in action: Exploring organizational culture. In P. Reason (Ed.), *Human Inquiry in Action*. London: Sage, pp. 199–220

Matenga, T. F. L., Zulu, J. M., Corbin, J. H. and Mweemba, O. (2021) Dismantling historical power inequality through authentic health research collaboration: Southern partners' aspirations. *Global Public Health* 16(1): 48–59.

Mbembe, A. (2002) African modes of self-writing. *Public Culture* 14(1): 239–273.

Melber, H. (2017) Knowledge production, ownership and the power of definition: Perspectives on and from Sub-Saharan Africa. In I. Baud, E. Basile, T. Kontinen and S. von Itter (Eds.), *Building Development Studies for the New Millennium*. London: Palgrave Macmillan, pp. 265–287.

Mkandawire, T. (1999) Social science and democracy: Debates in Africa. *African Sociological Review* 3(1): 20–34.

Mohan, G. and Wilson, G. (2005) The antagonist relevance of development studies. *Progress in Development Studies* 5(4): 261–278.

Mudimbe, V. Y. (1988) *The Invention of Africa: Gnosis, Philosophy and the Order of Knowledge*. Bloomington and Indianapolis: Indiana University Press.

Murphy, J. and Zhu, J. (2012) Neo-colonialism in the academy? Anglo-American domination in management journals. *Organization* 19: 915–927.

Nhemachena, A., Mlambo, N. and Kaundjua, M. (2016) The notion of the "field" and the practices of researching and writing Africa: Towards decolonial praxis. *The Journal of Pan African Studies* 9(7): 15–36.

Nkomo, S. M. (2011) A postcolonial and anti- colonial reading of 'African' leadership and management in organization studies: Tensions, contradictions and possibilities. *Organization* 18: 365–386.

Okere, T., Njoku, C. and Devisch, R. (2005) All knowledge is first of all local knowledge: An introduction. *Africa Development* 30(3). https://doi.org/10.4314/ad.v30i3.22226

Parker, M. and Kingori, P. (2016) Good and bad research collaborations: Researchers' views on science and ethics in Global Health Research. *PLoS ONE* 11(10): e0163579. https://doi.org/10.1371/journal.pone.0163579

Prasad, A. (2009) Contesting hegemony through genealogy: Foucault and cross cultural management research. *International Journal of Cross Cultural Management* 9(3): 359–369.

Publons (2018) *The Global State of Peer Review.* https://publons.com/static/Publons-Global-State-Of-Peer-Review-2018.pdf

Quayson, A. (1997) Protocols of representation and the problems of constituting an African 'Gnosis': Achebe and Okri. The yearbook of English studies. *The Politics of Postcolonial Criticism* 27: 137–149.

Reason, P. and Bradbury, H. (2001) Introduction: Inquiry and participation in search of a world worthy of human aspiration. In P. Reason and H. Bradbury (Eds.), *Handbook of Action Research. Participative Inquiry and Practice.* London: Sage, pp. 1–14.

Rodriguez, J. and Freeman, K. J. (2016) 'Your focus on race is narrow and exclusive:' The derailment of anti-racist work through discourses of intersectionality and diversity. *Whiteness and Education* 1(1): 69–82.

Ruggunan, S. D. (2016) Decolonising management studies: A love story. *Acta Commercii* 16(2): 103–138. https://doi.org/10.4102/ac.v16i2.412

Said, E. W. (2003) *Orientalism.* London: Penguin Books.

Sartre, J.-P. (1965) Black Orpheus. *The Massachusetts Review* 6(1): 13–52.

Shivji, I. G. (2006) Pan-Africanism or imperialism? Unity and struggle towards a new democratic Africa. *African Sociological Review* 10(1): 208–220.

Smith, L. T. (1999) *Decolonizing Methodologies. Research and Indigenous People.* London: Zed Books.

Spivak, G. C. (2006) *In Other Worlds. Essays in Cultural Politics.* New York: Routledge.

Sukarieh, M. and Tannock, S. (2019) Subcontracting academia: Alienation, exploitation and disillusionment in the UK overseas Syrian refugee research industry. *Antipode* 5(1): 664–680.

Taylor, H. (2001) Insights into participation from critical management and labour process perspectives. In B. Cooke and U. Kothari (Eds.), *Participation: The New Tiranny?* London: Zed Books, pp. 122–138.

Tinker, C. and Armstrong, N. (2008) From the outside looking in: How an awareness of difference can benefit the Qualitative Research process. *The Qualitative Report* 13(1): 53–69.

Turner, S. (2010) The silenced assistant: Reflections of invisible interpreters and research assistants. *Asia Pacific Viewpoint* 51(2): 206–219.

Wiegratz, J. (2009) *Uganda's Human Resource Challenge: Training, Business Culture and Economic Development.* Kampala: Fountain Publishers.

Wright, H. K. (2002) Notes on the (Im)possibility of articulating continental African identity. *Critical Arts* 16(2): 1–18.

Wray-Bliss, E. (2003) Research subjects/research subjections: exploring the ethics and politics of critical research. *Organization* 10(2): 307–325.

Zeleza, P. T. (2009) African studies and universities since independence. *Transition* 101: 110–135.

3

A DECOLONIAL FEMINIST ETHNOGRAPHY

Empowerment, ethics and epistemology

Jennifer Manning

Introduction

A decolonial feminist ethnography is an empowering research methodology that can situate the knowledge, lived experiences and worldviews of 'others' who are often marginalised in management research, thought and practice. This methodology focuses on the importance of ethics and epistemology in shaping the methods of knowledge production while striving for empowerment in the research process. A decolonial feminist ethnography is a messy, bricolaged way of doing research. It is also an empowering methodology that draws attention to differences, inequalities and 'otherness'. Reconfiguring critical ethnography to recognise the coloniality of power, a decolonial feminist ethnography enables researchers to consider and address the ethical and political implications of research and knowledge production.

I developed a decolonial feminist ethnography when undertaking my doctoral research. My research explored the work and lives of marginalised, indigenous Maya women working together in backstrap weaving groups in rural, remote Highland communities in Guatemala. Having previously lived in Guatemala while spending two formative years living, working and travelling through East Asia, the South Pacific, Central America and the Caribbean, I was profoundly impacted by how much I learnt from the various people and cultures throughout this experience. From complex worldviews, to challenging lived experiences, to alternative ways of working and organising, I came to a realisation that my worldviews were built upon an ontology of modernity that did not adequately recognise the work, lives and knowledge of 'others'. There is limited empirical engagement *with* marginalised, indigenous women in the Global South within management and organisation studies. Located outside the dominant Western discourse, little is known about how they construct their identity and their work/organisational experiences. Traditional ethnographies reinforce imperialist tendencies and epistemic violence, and

DOI: 10.4324/9780429352492-3

produce authoritative, descriptive studies about 'others' (Foley, 2002; Madison, 2012). However, from my experiences, I found that it is only through dialogue, which requires listening as much as talking, we can advance mutual understanding.

Developing a decolonial feminist ethnography enabled me, a white, European (Irish) woman, to engage with the politics of power and positionality in the research process so as to create space for marginalised Maya women to voice their own understanding of gender, identity and work from within the context of their social, cultural and historical location. This approach to research encourages researchers to strive towards being ethically and reflexively engaged throughout the research process. I continually tried to be ethically mindful and reflexive in my engagements with the Maya women participants in order to understand how I experienced our relationship and to know how to (re)present the women and their knowledge. Thus, a decolonial feminist ethnography highlights the need to consider deeply the personal, political and ethical considerations of research. I used the embodied performances of listening and moving to address the politics of power and positionality inherently embedded in the research process and to continually try to ensure I was ethically and reflexively engaged with the Maya women participants.

This chapter will first provide insight into the theory underpinning this methodological approach. I start with a brief theoretical overview of decolonial feminist theory and then draw out its relationship to critical ethnography. I discuss *how* a decolonial feminist ethnography can produce different forms of knowledge/ways of knowing in management and organisation studies by engaging in research *with* those who are often 'othered' and left in the margins of management thought and practice. Following this, moving beyond the theory, I explore the *doing* of decolonial feminist research. By drawing on my doctoral research experience, I share how I used the embodied performances of listening and moving to engage in a dialogue with the Maya women participants where power was shared and knowledge produced together. I then close this chapter by drawing out how a decolonial feminist ethnography is an empowering methodology and can contribute to the growing discourse on empowering ways of doing research.

In undertaking research and producing knowledge *with* 'others' in the Global South,[1] researchers need to continually try to ensure that they are not implicit in perpetuating the conditions of inequality or power domination in the research process, or the potential silencing of participants knowledge and voice, and reproducing their 'otherness'. This chapter makes a contribution to management research by providing insight, in both theory and practice, of an alternative, empowering way of undertaking research that is underpinned by an ethical commitment to participants by means of decolonising ourselves and the research process.

Understanding decolonial feminist theory

The theoretical and epistemological origins of decolonial feminism provide insights into how a decolonial feminist ethnography is an empowering methodology that can produce different forms of knowledge/ways of knowing. First, beginning with an introduction to decolonial theory, which is founded in the modernity/coloniality dialogues between prominent Latin American scholars. The work of Mignolo (2007, 2009, 2011), Escobar (2007, 2010), Dussel (Dussel and Ibarra-Colado, 2006) and Quijano (2000, 2007), among others, founded decolonial theory by critiquing Eurocentric modernity and claims of universality. Dussel and Ibarra-Colado (2006) explain modernity as a phenomenon that denotes the sociocultural centrality of Europe from the moment the Americas were discovered. Modernity refers to the crystallisation of discourses, practices and institutions that have developed over the past few hundred years from European ontological and cultural colonisation (Ceci Misoczky, 2011; Escobar, 2010). Escobar (2010) explains that the world and all knowledges constructed on the basis of an ontology of modernity became universal, and this universal ontology has gained dominance over certain worldviews, institutions, constructs and practices. Decolonial theorists argue that the idea of the universality of a Western ontology is based on the displacement of those in the Global South from the effective history of modernity. As a result, history becomes a product of the West, and modernity became synonymous with the West by displacing the actions, ideas and knowledge of those in the Global South. In so doing, Western modernity created the 'other'. The 'other' are those who do not fit the profile of modernity, that is, persons and cultures that are considered non-modern. The postcolonial theorist Spivak (1988) uses the term 'subaltern' to emphasise the position of the marginalised 'other', which refers to those socially, politically and geographically outside the dominant power structures.

Decolonial theorist Quijano (2000, 2007) developed the coloniality of power concept, which helps us understand how the knowledge, lived experiences and worldviews of the 'other' remain in the margins. The coloniality of power is the interrelation of four domains of power and control: control of economy (e.g., land appropriation, exploitation of labour, control of natural resources), control of authority (e.g., government, institution, army), control of gender and sexuality (e.g., family, education) and control of subjectivity and knowledge (e.g., epistemology, education and formation of subjectivity). Quijano (2007) argues that the coloniality of power is the persistent categorical and discriminatory discourse that is reflected in the social and economic structures of modern post-colonial societies. The coloniality of power simultaneously dismantles 'other' knowledges, social organisation and ways of life (Mignolo, 2007). Modernity and coloniality are mutually dependent phenomena; coloniality refers to 'the pattern of power which has emerged as a result of colonialism' and is an explicit strategy of epistemological control and domination (Ceci Misoczky, 2011, 347). As explained by Mignolo (2007, 162), 'modernity, capitalism and coloniality are aspects of the same package of control of economy and authority, of gender and sexuality, of knowledge and subjectivity'. Coloniality/modernity has created the culturally, socio-economically and politically marginalised 'other' of the Global South. As a result, the knowledge and practice of the 'other' remain in the margins of the social sciences, and management and organisation studies in particular.

Decolonial theory is thinking that emerges from and within the margins (Ibarra-Colado, 2006). This theoretical perspective calls for the decolonisation of knowledge so the epistemologies of those in the Global South, particularly those with subalternised racial/ethnic/religious/gendered spaces and bodies, can be taken seriously and moved from the periphery (Grosfoguel, 2007). In summary, decolonial theory can be understood as broadening non-Western modes of thought and ways of 'seeing and doing' while simultaneously demanding the acceptance of marginalised, different and alternative ontologies, epistemologies and worldviews (Escobar, 2007), or as put by (Bhambra, 2014, 120):

> Decoloniality [is] only made necessary as a consequence of the depredations of colonialism, but in [its] intellectual resistance to associated forms of epistemological dominance [it] offers more than simple opposition. [Decolonial theory] offers . . . the possibility of a new geopolitics of knowledge.

Understanding decolonial theory was my first step in implementing a decolonial feminist ethnography as it enabled me to understand how the knowledge, lived experiences and worldviews of the 'other' have been marginalised in mainstream academic discourse. However, as my research was engaging with the lived experiences of marginalised 'other' women, it was important to both me and the Maya women participants to integrate a feminist perspective. Thus, my next step was to explore decolonial feminist theory and integrate this into the development of my ethnography.

Few of the founders of decolonial theory directly acknowledge gender (Harding, 2016; Lugones, 2008, 2010; Paludi et al., 2019). Decolonial feminism is an emerging theoretical concept led by Lugones (2008, 2010) that centres on decolonial theory in racial/gendered feminist context. Decolonial feminism engages with debates pertaining to coloniality/modernity and Global South indigenous identity and gender, while also providing a space for the voices and experiences of marginalised 'other' women (Bhambra, 2014; Lugones, 2010; Schiwy, 2007). Much research 'explains' women as if the reality of White, Western, middle-class women applies to women from all cultures, classes, races and religions of the world (Anzaldúa, 2007; Mohanty, 1988, 2003; Parpart, 1993). Limited empirical engagement with marginalised women in the Global South perpetuates their 'otherness', and little is known about marginalised, indigenous, poor, Black/Brown, non-Westernised women: their voices are eclipsed by discourses *about* them (Espinosa Miñoso et al., 2014). Decolonial feminist theorising seeks to provide space for the silenced voices of women to speak of their identities, lived experiences and worldviews.

Decolonial feminist theory and critical ethnography

Now, with this brief synoptic understanding of decolonial feminism, I shift the context towards the transitioning of this theoretical lens into an empowering methodology. I integrate decolonial feminist theory into critical ethnography to create a decolonial feminist approach to research. Critical ethnography provides space to produce rich accounts of the field, as well the space to engage with the voices, perspectives and narratives of those who have been marginalised (Foley, 2002; Guba and Lincoln, 1994; Kincheloe, 2001; Kincheloe and McLaren, 2005; Madison, 2012; Till, 2009), while creating a dialogical relationship between the researcher and participants (Foley, 2002; Madison, 2012) by fostering conversation and reflection (Kincheloe and McLaren, 2005). Understanding the 'other' is one of the primary motivations for doing ethnographic research (Krumer-Nevo and Sidi, 2012); however, the desire of researchers to know the 'other' and invite them to speak is a potential source of dominance (Manning, 2018). In the early stages of developing my decolonial feminist ethnography, I spent much time reflecting on the question posed by Kincheloe et al. (2015, 171): 'How can researchers respect the perspective of the "other" and invite the "other" to speak?'. To address this complexity, I needed a methodology that moved beyond a well-intentioned critical ethnography towards one that would encourage me at all times to accept the knowledge, worldviews and lived experiences of the 'other' without imposing a Western ontology of modernity, while also enabling me to engage with 'others' without perpetuating their 'otherness'.

To this end, I integrated the epistemology of decolonial feminist theory into the methodology of critical ethnography. This is a bricolaged approach to research which understands research as an eclectic process that takes place in a complex lived world and positions the researcher in the research process (Denzin and Lincoln, 1999). A decolonial feminist ethnography explicitly encourages the researcher to acknowledge and address the politics of positionality and power. The researchers' identity, position, privilege and power in fieldwork affect all aspects of the research process. In traditional ethnographies the researcher is in a position of power in relation to the knowledge that is produced, the representation of participation and the participant's knowledge. Thus, power relations are embedded in ethnography, which can produce imperialist tendencies in representing participants and their knowledge and thereby collude with structures of domination (Fine, 1994; Manning, 2018; Said, 1993). Western academics often claim epistemological authority over the 'other' by suggesting that they must be represented as they cannot represent themselves (Manning, 2018; Said, 1978; Spivak, 1988). As a result, researchers claim to know and speak for the 'other' and take ownership of the knowledge produced.

Integrating decolonial theory into a critical ethnography enables us to open up space 'for the reconstruction and the restitution of silenced histories, repressed subjectivities, subalternized knowledge and languages' and emphasises the need for the de-coloniality of power and knowledge (Mignolo, 2007, 451; Ceci Misoczky, 2019; Quijano, 2000, 2007). A decolonial feminist ethnography can help achieve epistemological decolonisation by enabling researchers contribute to social liberation by engaging in research that challenges inequality and domination in the research process (Ceci Misoczky, 2019; Quijano, 2007). This research approach advocates for researchers to consider the political and ethical implications of their research and encourages a dialogic performance between researcher and participants where power is shared and knowledge produced together (Manning, 2018). This is what decolonial feminist Lugones (1987, 637) refers to as 'world travelling' and 'loving perception':

> The reason why I think that travelling to someone's 'world' is a way of identifying with them is because by travelling to their 'world' we can understand what it is to be them [sic] and what it is to be ourselves in their eyes. Without knowing the other's 'world', one does not know the other, and without knowing the other one is really alone in the other's presence because the other is only dimly present to one. Through travelling to other people's 'worlds' we discover that there are 'worlds' in which those who are the victim of arrogant perception are really subjects, lively beings, resisters, constructors of visions even though in the mainstream construction they are animated only by the arrogant perceived and are pliable, foldable, file-awayable, classifiable.

To collate, a decolonial feminist ethnography is a bricolage approach to research that asserts an understanding that 'the positioning of the researcher in the social web of reality is essential to the production of rigorous and textured knowledge' (Kincheloe, 2005, 119). In this way, bricolage encourages decolonial feminist ethnographers to address the complexities of the lived world and the complications of power (Kincheloe and McLaren, 2005), thereby enabling researchers to better conceptualise the complexity of the research act (Denzin and Lincoln, 1999; Kincheloe, 2001). Thus, in addressing the complications of power and exploring the ways power shapes knowledge, researchers can embrace 'loving perception' and 'world travelling', whereby the different worldviews, knowledges and lived experiences of 'others' can be understood and explored together (Kincheloe and McLaren, 2005; Manning, 2018; Ceci Misoczky, 2019). By helping researchers understand the ways the coloniality of power influences the social, cultural, gendered, historical, economic and political conditions under which knowledge is produced, a decolonial feminist ethnography works to dismantle mainstream thinking and practice that, perhaps unknowingly, are implicit in perpetuating 'otherness' through the reproduction of systems of class, race and gender oppression, and encourages the questioning of dominant systems and knowledges (Kincheloe et al., 2015; Manning, 2018; Mignolo, 2007; Ceci Misoczky, 2019).

Undertaking a decolonial feminist ethnography in practice and performance

In theory, a decolonial feminist ethnography is an empowering, ethically engaged methodology that can challenge the coloniality of power to bring forward different worldviews, knowledges and lived experiences. However, in practice, the undertaking of this research approach is complex and challenging. To address the politics of power and positionality when engaging in research with the 'other' a decolonial feminist ethnographer must engage in self-reflexivity throughout the research process and explicitly explore power relations and representational practices (who produces and owns knowledge) (Brewis and Wray-Bliss, 2008; Manning, 2016, 2018; Őzkazanç-Pan, 2012). In undertaking my research, I had to question how, and if, I can represent the lived experiences of marginalised Maya women and encourage myself to openly confront the issues of power and ethics in my research.

The role of reflexivity

Integrating decolonial feminist thought into ethnographic practice requires deep engagement with how the self is involved in the ethnographic research process. As such, the first step in the practice of a decolonial feminist ethnography is to engage in reflexive practices. Reflexivity questions our relationship with our social world and the way in which we understand our experiences (Cunliffe, 2003). Being reflexive encourages us to be honest in the motivations that bring us to our research and also to be honest about our identities, positions, power, assumptions and so on, when engaging in research (Alvesson et al., 2008; Cunliffe and Karunanayake, 2013; Hardy et al., 2001; McDonald, 2013). A decolonial feminist ethnography embraces reflexivity, and reflexive practices encourage the questioning of the researchers position as (re)presenter of participants and their knowledge, the examination of power relationships and the recognition of the intersections of voice, place and privilege throughout the research process (Őzkazanç-Pan, 2012; Sultana, 2007).

The many social, economic and cultural differences between myself and the Maya women placed me in an irreconcilable position of difference, and I had to regard myself as the 'other' and reflexively question the situated, socially constructed nature of my self and my research participants (Foley, 2002). This is particularly important in the context of research with multiple axes of difference, inequalities and geopolitics (Manning, 2018; Sultana, 2007). There are clear ethnic, social and cultural dichotomies of privileged-poor, educated-unschooled, rural-urban and White-Brown that greatly influenced my relationship with the women participants. I was never going to be able to remove the physical, economic and social differences between the women and me. My identity and knowledge are formed within a European ontology of modernity, and I experience our world as a White woman whose Western culture and epistemology are considered transcendent above all others. The Maya women participants experience our world in a vastly different way. Theirs is one where their gender, culture, experiences and knowledge are regarded as inferior and remain absent from a universalised, Western modernity. As a result, I had to challenge my thinking, embrace our differences and seek out ways to form a commonality to overcome my difference and position of power.

To address this, I wrote ongoing diary-like notes examining myself, including my expectations, assumptions, bias, power and so on, unpacking how my relationship with the women participants was evolving and questioning the ways in which

I de-centre myself from the research to ensure the women's voices and knowledge were emerging through the research. Reflexivity is equally important after field-work. As researchers we are in control of analysing the data and presenting the findings, and thus in a position of power over the participants. During data analysis, I continually questioned what I was seeing and why I was seeing it. I wrote notes about themes that were emerging, and alongside this, I wrote further notes about the context in which themes emerged in the field. Providing context to emerging findings helps resists 'othering' by developing narratives that reflect the women's social, cultural and historical location (Krumer-Nevo and Sidi, 2012).

By engaging and articulating the politics of my power and positionality, I was able to support the production of knowledge that is located in the lived experiences of the marginalised 'other'. Continually exploring and questioning our relationship, embracing differences, contextualising the research and questioning my authority to represent the Maya women participants and their knowledge help to reduce privilege and distance, creating a more symmetrical power relationship between the self and 'other' (Fine, 1994; Krumer-Nevo and Sidi, 2012). The research-er's level of reflexivity and the choices we make during data collection, analysis and writing help disrupt traditional power imbalances which often dominate the research process. Empowerment in a decolonial ethnography comes from the space created between the researcher and participants where participants are agentic in the research process and the terms of the researcher-participant relationship. In my decolonial feminist ethnography the practice of creating an empowering meth-odology with more symmetrical power relations between myself and the Maya women participants emerged from my reflective practices through to my ethno-graphic performances of moving and listening.

The ethnographic performances of moving and listening

The *doing* of a decolonial feminist ethnography is a messy, non-linear, improvisa-tional methodology. It can be enacted differently within different research contexts. It was only when I entered the field that I was able to understand how I could build relationships with the women participants and create an empowering space. Our relationship was more than a dichotomy of insider-outsider and sameness-differ-ence; it was a space where power was explored and knowledge produced through moving and listening during the preformed ethnography. Preformed ethnographies simultaneously help build relationships and gather data using a range of methods and performances. This bricolage approach embraces a diversity of data collec-tion methods through improvisation in the field (Madison, 2012). Various forms and combinations of interviews, dialogue, field notes, stories, newspaper articles, historic documents, digital imagery, movement, literary texts and workshops are all brought together and improvised throughout the ethnographic performance (Denzin, 2003; Madison, 2012, 2018).

While engaging in reflexive practices, the sensory performances of moving and listening emerged as key methods during my decolonial feminist ethnography.

Together with this, my gender became integrated into my ethnographic performance. As a woman, I could participate in their traditional gendered division of labour that occupied a considerable amount of their time, enabling me to build relationships and reduce power distance. I first gained permission from the women to participate in the everydayness of their work and lives, and for three months, I spent from early morning into the evening with the women in their rural homes in their sparsely populated remote communities. During this time, I moved with them. There is fluidity and flexibility in the women's work and home lives. The women's everydayness consisted of cooking, cleaning, weaving, maintaining livestock and agriculture and meeting with each other to discuss matters relating to their weavings and product orders. To de-centre myself and position of power, and to build relationships with the women participants, I was actively engaged in their daily lives. I helped prepare meals; peeled, cleaned and ground corn (their dietary staple eaten with every meal); washed dishes; organised weaving materials and thread; sat with the women as they strapped themselves into their backstrap weave; cared for livestock; and followed them into their agriculture fields to maintain and harvest crops. Differences remained between us, but little actions that may even seem mundane can be significant in building relationships. The women grew comfortable with me in their homes, their lives and their work. I would always be the 'other' in our relationship, but moving with the women and participating in their everydayness not only enabled me to build relationships and gather rich data, it facilitated the decentring of myself from the research process. This is constitutive to a decolonial feminist ethnography as it is the knowledge, experiences and worldviews of the 'other' that emerge from the research. The researcher is the medium through which these new geopolitics of knowledge emerge.

The environment in which the research takes place shapes relationships and can have a profound impact on the dialogue and emerging data (Evans and Jones, 2011; Kusenbach, 2003). In moving with the Maya women participants, I was able to de-centre myself and create a more informal environment. This shifts the focus for participants; they are not concentrating on providing the 'correct' answer in an interview but engaged in an informal discussion in a relaxed atmosphere. Together we were exposed to the multi-sensory stimulation of the surrounding environment; animals roaming the open family compound, children playing, family members coming and going, all within a high altitude tropical, luscious green landscape with their homes concealed by large swaths of cornfields. Engaging with the women in their home environment offers privileged insight into both place and self in a more intimate setting. This enabled me to enhance my understanding of the women's relationship to their environment within the context of their gendered identities and social and cultural location. Movement in a performed ethnography helps the researcher to understand the lived experience of the 'other' and provides context to the knowledge being produced, and, in so doing, helps the researcher address the politics of power and positionality.

Moving and talking with the women throughout the day, on their terms and in their homes and communities, enabled the women to go about the everydayness of their work and lives. I was the 'other' trying to understand their world and to do this I had to enter their world on their terms. I adapted to the women's lives and moved fluidly with them through their daily lives, wherever they went I went and whatever they did I did. Much of our relationship building and dialogue emerged during our informal conversations as we prepared food. Mealtime is central to the familial social experience in the Maya women's homes, and it was during this time that the women talked freely; we shared experiences and talked about our lives. Together we created a casual, relaxed environment. I helped prepare food and peel vegetables, and the women laughed at how poorly I made tortillas. These informal interactions and dialogue became ongoing negotiated spaces for the development of symmetrical power relations, as well as reflexive identity construction and our relationship development. The performances of moving and listening became the vehicles through which I engaged in Lugones' (1987) concepts of 'world travelling' and 'loving perception'. Performances enable researchers to travel to the worlds of the 'other' and enter domains of intersubjectivity that problematise how we categorise ourselves and the 'other' and how we see ourselves through the 'others' eyes (Lugones, 1987; Madison, 2018). 'Loving perception' requires researchers to use space and dialogue to diffuse power and authority when 'world travelling'.

The sensory performance of movement generates richer data by being able to understand the connectedness of the women to their environment and also capture distinctive characteristics of place (Evans and Jones, 2011), while embodied expressions help to understand experiences (Conquergood, 2002). Moreover, it decentres the authority of the researcher and situates knowledge production in a particular time and place. Conquergood (2002) argues that the sensory practice of bodily movement is a democratic, ethical endeavour. It contrasts the mediations of distance and detachment to an embodied mode of aliveness in interactive engagement and togetherness with the 'other' on intersubjective ground.

Through reflexivity, movement and dialogue, I was able to build collaborative relationships and strive for the empowerment of the Maya women participants in the research process. Madison (2012, 186) explains that it is through dialogue and listening that researchers resist the process of 'othering', as dialogue embraces 'diversity, difference, and pluralism'. Citing Conquergood (1985), Madison (2012) draws out the relationship between reflexivity and listening; listening invites dialogue and dialogue encourages reflection on relationships and the tensions between the self and 'other' in the research process. The result of this brings many different voices into the research without anyone silencing the other. Having a dialogue with participants moves them into the research process (as opposed to being objects of the research) as they are involved and engaged in the production of knowledge. Traditional ethnographies can produce imperialist tendencies in representing participants and their knowledge and thereby collude with the structures of domination (Fine, 1994; Said, 1993; Tuhiwai Smith, 1999). A dialogue with participants is vital to decolonial ethnography as power relations are embedded in ethnography. Dialogic performance helps the researcher address politics/power relations and representational practices and facilitates the mutual creation of knowledge. It is part of 'world travelling', whereby through dialogue the researcher and participant engage in the co-performance of knowledge production. Dialogic performance becomes the space for the 'other' to engage equally in the research process and the researcher to become ethically invested in collaborative representation (Madison, 2006).

Through dialogue the Maya women participants become partners and agents in the research process. Together we explored different areas of their work and lives to workout meaning together. In the practice of my decolonial feminist ethnography, this involved transcribing and reviewing of all interviews, conversations, field notes, observations, photographs and various documents at the end of each day. This enabled me to develop a preliminary understanding of the women's lives and work. With this, the following day I discussed the development of my understanding with the women to gauge their opinion and perspective. In so doing, provisional ideas and findings were worked out together. This became an iterative, ongoing process throughout my research. The women were agentic in the research process, thereby ensuring that I was not re-imposing dominant structures that have oppressed them but acknowledging their agency in the telling of their story.

The performance of listening in a decolonial feminist ethnography goes beyond the sense of simply hearing. Listening is a multi-layered, multi-sensory engagement, whereby relationships are built through connectedness and collaboration, instilling an ethics of co-creation in the research process (Madison, 2018; Fischlin et al., 2013). As explained by Madison (2012), listening goes beyond participants being heard and included, but focuses on voice. Voice is the embodied self of the participants constructed by their social, cultural and historical experiences. Voice, in a decolonial feminist ethnography, recognises subjectivity. Thus, through listening to participants, their personal experiences, knowledge, struggles, resilience, cultural politics etc., are engaged with and explored. The researcher, symbolically and temporarily, enters into participants' locations of voice within their own experiences (Madison, 2012). Participants' voices, perspectives and narratives emerge from listening, and driven by an ethical responsibility to their participants, researchers communicate their worlds in their own words. Although the researcher is writing and representing the participants' narratives, the narratives reflect their socially situated lived experiences. In the context of my research, I had to acknowledge that I write from a position of power and privilege, but at the same time, the voices of the Maya women are located with mine in the telling of their story. My decolonial feminist ethnography was localised and grounded in the Maya women's meaning of themselves and their work, and through reflection, listening and moving, we explored their work and lives together. A decolonial feminist ethnography enables the voice of research participants to be heard. However, this voice is not a romanticised representation of the 'other' nor a perpetuation of their 'otherness', but a representation of their lived experiences grounded in their subjectivities.

A decolonial feminist ethnography is a bricolage approach to critical research that embraces the improvisations of sensory preformed ethnography. Together with this, the power and politics of a decolonial feminist ethnography demand ethical responsibility on behalf of the researcher. The coloniality of power has marginalised the voice, knowledge and subjectivities of the 'other'. As put by Conquergood (2002, 146), 'what gets squeezed out by this epistemic violence is the whole realm of complex, finely nuanced meaning that is embodied, tacit, intoned, gestured, improvised, co-experienced'. To strive for the empowerment of research participants, a decolonial feminist ethnography is built on an ethical commitment to respect the voice and knowledge of the 'other' and recognises the politics of power and positionality embedded in the research process. The sensory performances of moving and listening facilitate this by embracing the co-production and co-ownership of knowledge and bring forward different worldviews, knowledges and lived experiences. Removing the abstract and authoritative study of/about subjects, sensory performances provide an opportunity for 'another way of knowing that is grounded in active, intimate, hands-on participation and personal connection' *with* the 'other' (Conquergood, 2002, 146). Integrating moving and listening into my decolonial feminist ethnography helped me build relationships and address power differentials between myself and the Maya women participants by paying attention to the issues of voice, interactions and dialogue to ensure that they were agentic in the production of knowledge about them.

Conclusion

This ethnographic approach is just one of many empowering methodologies that create space for more engaged social and organisational research practice (e.g., Bell and Willmott, 2020; Cunliffe and Karunanayake, 2013; Cunliffe and Scaratti, 2017; Reedy and King, 2019; Madison, 2018; Özkazanç-Pan, 2012, among others). This chapter, and the many empowering methodologies challenging traditional research practices, has argued that it is vital to create space for more situated approaches to producing knowledge in management research that are grounded in ethics and epistemology, as opposed to Western 'scientific rigour', whereby a Western ontology of modernity dominates academic agendas and neo-colonial relations shape the practice of knowledge production (see Bell et al. 2017). Empowering methodologies foster greater pluralism and innovative approaches to knowledge production, which can help to pave the way towards the eradication of core-periphery relations in academic knowledge production and dissemination. This is a movement grounded in researcher responsibility and participant agency. It emphasises the need for researchers to take responsibility on how knowledge is produced and the impact of our knowledge claims. This chapter makes a contribution to this movement by providing insight into an alternative empowering approach to research. I have demonstrated how a decolonial feminist ethnography can disrupt the politics of power and positionality and explored why this is important. Specifically, this chapter contributes insight into: (1) the theoretical development of a decolonial feminist ethnography, and (2) how this research can be performed in practice to produce knowledge *with* the 'other'.

Engaging in reflexivity and addressing the power and politics in knowledge production it become possible to challenge the coloniality of power and integrate the voices, lived experiences and worldviews of 'others' into mainstream management and organisation studies, and thereby produce different forms of knowledge/

ways of knowing and create a new geopolitics of knowledge. Empowering methodologies have the ability to promote social transformation. By challenging inequalities that are embedded in the research process, particularly with those who are 'othered', an empowering methodology ensures participants are agents in the production of knowledge about them (Davis, 2008; Ross, 2017). A decolonial feminist ethnography is an empowering methodology that acknowledges the researcher's responsibility to research participants, particularly the 'other', and addresses this in practice by means of ongoing reflexivity to ensure participants are agentic in research. This approach to research thereby focuses on the importance of ethics and epistemology in shaping the processes of knowledge production.

'Loving perception', dialogic performance and ethics are among the many concepts and practices integrated into my decolonial feminist ethnography, yet they all require researchers to understand that all individuals, including their knowledge, worldview and lived experiences, are valuable and deserving of understanding. This in turn leads to a more understanding society where differences and 'otherness' are accepted, while also offering a 'more adequate, richer, better account of a world, in order to live in it well and in a critical, reflexive relation to our own and others' practices' (Haraway, 1988, 579).

In undertaking a decolonial feminist ethnographic performance, researchers can come to understand the worldviews and the lived experiences of the 'other'. This is an act of integrating ethics, epistemology and empowerment. Decolonial feminist theory critiques Western representation of the 'other' and reveals how knowledge produced in and by the West is layered with colonial power, thereby creating and sustaining a politics of Western knowledge dominance and rendering the 'other' an object of knowledge (Mignolo, 2007; Prasad, 2003; Said, 1978). Applying this theoretical lens to critical ethnography enables a researcher to understand knowledge as situated. That is, knowledge is embedded within a social, cultural, historical and political time and place that reflects contextual features and lived experiences (Haraway, 1988). A decolonial feminist ethnography values all knowledge and lived experiences as equal, and in so doing provides a new framework within the geopolitics of knowledge production, one that demands respect for the pluralisation of differences.

Note

1 The 'Global South' is a highly politically contested and debated discourse. It refers to the geographic, socio-economic and political divide that exists between the countries of the economically 'developed world', known as the Global North or the West, and the countries that are referred to as 'Third World' or 'developing nations', primarily former colonies of the Global North that are seen as poor (Prashad, 2012). I use the term 'Global South' throughout this chapter to refer to the countries that are victims of, firstly, colonisation and, subsequently, capitalist mal-development, and, as such, they are considered economically developing or underdeveloped.

References

Alvesson, M., Hardy, C. and Harley, B. (2008) Reflecting on reflexivity: Reflexive textual practices in organisation and management theory. *Journal of Management Studies* 45(3): 480–501.

Anzaldúa, G. (2007) *Borderlands: The New Mestiza.* San Francisco, CA: Aunt Lute Books.

Bell, E., Kothiyal, N. and Willmott, H. (2017) Methodology-as-technique and the meaning of rigour in globalized management research. *British Journal of Management* 28(3): 534–550.

Bell, E. and Willmott, H. (2020) Ethics, politics and embodied imagination in crafting scientific knowledge. *Human Relations* 73(10): 1366–1387.

Bhambra, G.K. (2014) Postcolonial and decolonial dialogues. *Postcolonial Studies* 17(2): 115–121.

Brewis, J. and Wray-Bliss, E. (2008) Re-searching ethics: Towards a more reflexive critical management studies. *Organization Studies* 29(12): 1521–1540.

Ceci Misoczky, M. (2011) World visions in dispute in contemporary Latin America: development x harmonic life. *Organization* 18(3): 345–363.

Ceci Misoczky, M. (2019) Contributions of Aníbal Quijano and Enrique Dussel for an anti-management perspective in defence of life. *Cuadernos de Administración* 32(58).

Conquergood, D. (1985) Performing as a moral act: Ethical dimensions of the ethnography of performance. *Text and Performance Quarterly* 5(2): 1–13.

Conquergood, D. (2002) Performance studies: Interventions and radical research. *The Drama Review* 46(2): 145–156.

Cunliffe, A.L. (2003) Reflexive inquiry in organizational research: Questions and possibilities. *Human Relations* 56(8): 983–1003.

Cunliffe, A.L. and Karunanayake, G. (2013) Working within hyphen-spaces in ethnographic research: Implications for research identities and practice. *Organizational Research Methods* 16(3): 364–392.

Cunliffe, A.L. and Scaratti, G. (2017) Embedding impact in engaged research: Developing socially useful knowledge through dialogical sensemaking. *British Journal of Management* 28(1): 29–44.

Davis, C.S. (2008) Empowerment. In *The Sage Encyclopaedia of Qualitative Research Methods,* edited by Lisa M. Given, 260–261. London: Sage.

Denzin, N.K. (2003) *Performance Ethnography: Critical Pedagogy and the Politics of Culture.* Thousand Oaks, CA: Sage Publications.

Denzin, N.K. and Lincoln, Y.S. (1999) *The Sage Handbook of Qualitative Research.* Thousand Oaks, CA: Sage Publications.

Dussel, E. and Ibarra-Colado, E. (2006) Globalization, organisation and the ethics of liberation. *Organisation* 13(4): 489–508.

Escobar, A. (2007) Worlds and knowledges otherwise: The Latin American modernity/coloniality research program. *Cultural Studies* 21(2–3): 179–210.

Escobar, A. (2010) Latin America at a crossroads. *Cultural Studies* 24(1): 1–65.

Evans, J. and Jones, P. (2011) The walking interview: Methodology, mobility and place. *Applied Geography* 31(2): 849–858.

Fine, M. (1994) Working the hyphens: Reinventing self and other in qualitative research. In *The Sage Handbook of Qualitative Research,* edited by Norman K. Denzin and Yvonna S. Lincoln, 70–82. Thousand Oaks, CA: Sage Publications.

Fischlin, D., Heble, A. and Lipsitz, G. (2013) *The Fierce Urgency of Now: Improvisation, Rights, and the Ethics of Cocreation.* Durham, NC: Duke University Press.

Foley, D.E. (2002) Critical ethnography: The reflexive turn. *International Journal of Qualitative Studies in Education* 15(4): 469–490.

Grosfoguel, R. (2007) The epistemic decolonial turn. *Cultural Studies* 21(2–3): 211–223.

Guba, E.G. and Lincoln, Y.S. (1994) Competing paradigms in qualitative research. In *The Handbook of Qualitative Research*, edited by Norman K. Denzin and Yvonna S. Lincoln, 105–117. Thousand Oaks, CA: Sage Publications.

Haraway, D. (1988) Situated knowledges: The science question in feminism and the privilege of partial perspective. *Feminist Studies* 14(3): 575–599.

Harding, S. (2016) Latin American decolonial social studies of scientific knowledge: Alliances and tensions. *Science, Technology, & Human Values* 41(6): 1063–1087.

Hardy, C., Phillips, N. and Clegg, S. (2001) Reflexivity in organization and management theory: A study of the production of the research subject. *Human Relations* 54(5): 531–560.

Ibarra-Colado, E. (2006) Organization studies and epistemic coloniality in Latin America: Thinking Otherness from the margins. *Organization* 13(4): 463–488.

Kincheloe, J.L. (2001) Describing the bricolage: Conceptualizing a new rigor in qualitative research. *Qualitative Inquiry* 7(6): 679–692.

Kincheloe, J.L. (2005) *Critical Constructivism Primer*. New York, NY: P. Lang.

Kincheloe, J.L. and McLaren, P. (2005) Rethinking critical theory and qualitative research. In *The Sage Handbook of Qualitative Research*, edited by Norman K. Denzin and Yvonna S. Lincoln, 303–342. Thousand Oaks, CA: Sage Publications.

Kincheloe, J.L., McLaren, P. and Steinberg, S.R. (2015) Critical pedagogy and qualitative research: Moving to the bricolage. In *The Landscape of Qualitative Research*, edited by Norman K. Denzin and Yvonna S. Lincoln, 163–177. Thousand Oaks, CA: Sage Publications.

Krumer-Nevo, M. and Sidi, M. (2012) Writing against othering. *Qualitative Inquiry* 18(4): 299–309.

Kusenbach, M. (2003) Street phenomenology: The go-along as ethnographic research tool. *Ethnography* 4(3): 455–485.

Lugones, M. (1987) *Playfulness, World-Travelling, and Loving Perception. The Woman That I Am: The Literature and Culture of Contemporary Women of Color*. PhD Dissertation, State University of New York.

Lugones, M. (2008) The coloniality of gender. *Worlds & Knowledges Otherwise* 2(Spring): 1–17.

Lugones, M. (2010) Toward a decolonial feminism. *Hypatia: A Journal of Feminist Philosophy* 25(4): 742–759.

Madison, D.S. (2006) The dialogic performative in critical ethnography. *Text and Performance Quarterly* 26(4): 320–324.

Madison, D.S. (2012) *Critical Ethnography: Method, Ethics and Performance*. Thousand Oaks, CA: Sage Publications.

Madison, D.S. (2018) *Performed Ethnography & Communication: Improvisation and Embodied Experience*. Abingdon: Routledge.

Manning, J. (2016) Constructing a postcolonial feminist ethnography. *Journal of Organizational Ethnography* 5(2): 90–105.

Manning, J. (2018) Becoming a decolonial feminist ethnographer: Addressing the complexities of positionality and representation. *Management Learning* 49(3): 311–326.

McDonald, J. (2013) Coming out in the field: A queer reflexive account of shifting researcher identity. *Management Learning* 44(2): 127–143.

Mignolo, W.D. (2007) Introduction: Coloniality of power and de-colonial thinking. *Cultural Studies* 21(2): 155–167.

Mignolo, W.D. (2009) Epistemic disobedience, independent thought and decolonial freedom. *Theory, Culture & Society* 26(7–8): 159–181.

Mignolo, W.D. (2011) Geopolitics of sensing and knowing: On (de)coloniality, border thinking and epistemic disobedience. *Postcolonial Studies* 14(3): 273–283.

Miñoso, Y.E., Gómez Correal, D. and Ochoa Muñoz, K. (2014) *Tejiendo de otro Modo: Feminismo, epistemología, y apuestas decoloniales en Abya Yala*. Popayan, CO: Editorial Universidad del Cauca.

Mohanty, C. (1988) Under Western eyes: Feminist scholarship and colonial discourses. *Feminist Review* 30(1): 61–88.

Mohanty, C. (2003) *Feminism without Borders: Decolonizing Theory, Practicing Solidarity*. Durham: Duke University Press.

Őzkazanç-Pan, B. (2012) Postcolonial feminist research: Challenges and complexities. *Equality, Diversity and Inclusion: An International Journal* 31(5/6): 573–591.

Paludi, M.I., Helms Mills, J. and Mills, A. (2019) Cruzando fronteras: The contribution of a decolonial feminism in organization studies. *Management & Organizational History* 14(1): 55–78.

Parpart, J.L. (1993) Who is the 'other'? A postmodern feminist critique of women and development theory and practice. *Development and Change* 24(3): 439–464.

Prasad, A. (2003) *Postcolonial Theory and Organizational Analysis: A Critical Engagement*. New York: Palgrave Macmillan.

Prashad, V. (2012) Dream history of the Global South. *Interface: A Journal for and about Social Movements* 4(1): 43–53.

Quijano, A. (2000) Coloniality of power and eurocentrism in Latin America. *Nepantla: Views from South* 1(3): 533–580.

Quijano, A. (2007) Coloniality and modernity/rationality. *Cultural Studies* 21(2): 168–178.

Reedy, P.C. and King, D.R. (2019) Critical performativity in the field: Methodological principles for activist ethnographers. *Organizational Research Methods* 22(2): 564–589.

Ross, K. (2017) Making empowering choices: How methodology matters for empowering research participants. *Forum Qualitative Sozialforschung/Forum: Qualitative Social Research* 18(3).

Said, E.W. (1978) *Orientalism*. New York: Vintage Books.

Said, E.W. (1993) *Culture and Imperialism*. New York: Vintage Books.

Schiwy, F. (2007) Decolonization and the question of subjectivity: Gender, race and binary thinking. *Cultural Studies* 21(2–3): 271–294.

Spivak, G.C. (1988) Can the subaltern speak? In *Marxism and the Interpretation of Culture*, edited by C. Nelson and L. Grossberg, 271–313. Urbana: University of Illinois Press.

Sultana, F. (2007) Reflexivity, positionality and participatory ethics: Negotiating fieldwork dilemmas in international research. *ACME: An International E-Journal for Critical Geographies* 6(3): 374–385.

Till, K. (2009) Ethnography. In *International Encyclopaedia of Human Geography*, edited by R. Kitchin and N. Thrift, 626–631. Oxford: Elsevier.

Tuhiwai Smith, L.T. (1999) *Decolonizing Methodologies: Research and Indigenous Peoples*. London: Zed Books.

4

VULNERABILITY AS PRAXIS IN STUDYING SOCIAL SUFFERING

Devi Vijay

Introduction

Organisation studies have become increasingly attentive towards addressing complex social problems and grand societal challenges. Nascent research streams adopt a "public value" approach that embraces societal concerns and public dialogue (Delbridge 2014) to examine the organisational and institutional structures that produce, amplify, or mitigate social suffering. Social suffering here is the "assemblage of human problems" that arise from, and have consequences in, the political, economic, and institutional forces constituting human experience (Kleinman, Das and Lock 1997, ix). Social suffering as a concept destabilises boundaries such as health, moral, legal, human rights issues, to focus on the trauma, injuries, and pain that arise from complex social problems.

This chapter considers specific ethico-political imperatives underpinning studies of social suffering in organisational theory, and by extension, social theory. I then draw upon Judith Butler's scholarship to explore vulnerability as praxis. Like any sociological practice, studying social suffering wrestles with the arbitrary intrusions and inherent asymmetries inherent in research encounters, with the researcher setting up the rules of the game, making "private worlds public," opening up statements made in the context of a certain trust between individuals (Bourdieu 1999, 608). However, understanding social suffering is compounded by particular challenges.

First, those at risk of suffering can be prone to intrusive empiricism wherein lives are opened up for scrutiny, voyeuristic revelation, and fast and swift exposés (Back 2012). The consequence is a typecasting of lives in specific ways that come to characterise the researched. Accordingly, suffering can be essentialised, naturalised, or sentimentalised. Yet, suffering is contingent on time, space, communities, and values. Individuals do not suffer in the same way (Kleinman et al. 1997). Further, the swiftness of storying, the "trap of now" without attentiveness to a larger scale and longer historical time frame can occlude and obscure multiple, interconnected analytic registers of

DOI: 10.4324/9780429352492-4

suffering (Back 2012). While sensational journalism and reality TV typify this kind of intrusive empiricism, academic work, increasingly subject to neoliberal compulsions of productivity and efficiency, can be similarly fallible. Second, through a politics of pity, the spectator can create a "spectacle of suffering," commodifying suffering (Boltanski 1999). Such cultural appropriations of suffering are increasingly integral to media imaging for corporate profiteering and global marketing. For example, a Pulitzer Prize–winning *New York Times* photograph by Kevin Carter portrayed a vulture perched near a starving Sudanese child. Kleinman and Kleinman (1997) open up the moral and political assumptions inhered in such acts of moral witnessing. Similarly, inequality and social suffering are put to use in proliferating bottom-of-the-pyramid discourses which serve neoliberal interests. Here, the subaltern is the object of the suffering of the world, and also the object of rescue (Chatterjee 2020). Accordingly, social suffering may be alleviated through market-based solutions such as microfinance (anchored around women's self-help groups), which "mystify capitalism as a compassionate and caring system," while normalising privatisation (Khare and Varman 2017, 1596). A third challenge arises from the tension of representation inhered in hegemonic knowledge structures of capitalism, patriarchy, and colonialism (Vijay and Varman 2018; Vijay, Gupta and Kaushiva 2020). These structures of domination determine what we recognise as social suffering.

How then may we think ourselves out of these spaces of domination? How do we, located in elite institutions in the Global South, implicated in hegemonic knowledge production in the Global North, negotiate our contradictory subjectivities in field sites without reproducing structures of domination? How do we remain respectful to those who allow us into the intimacy – the suffering and pain, but also care and solidarity – of their worlds? How do we think through a praxis for social transformation that has the objective of transforming derealisation, objectification, and suffering? How do we write without appropriating suffering?

Reflexivity is the researcher's craft practice to examine her epistemological, ontological assumptions, and ethico-political positions (Bell and Willmott 2020; Krumer-Nevo and Sidi 2012). However, this reflexivity has its limits by centering the self with relation to the other (Mohanty 1984). Reflexivity can engage with the self as self-sufficient, intelligible, as "bounded and deliberate individuals, self-propelling and self-driven" (Butler and Athanasiou 2013, 4). Such reflexivity can well morph into "thinly veiled self-obsession" (Rhodes and Carlsen 2018, 1300). The researcher can declare oneself as white/male without rupturing racial, class, colonial, or gender privilege (Swan 2017). In other words, by admission of oneself as white male, one may recognise racism and patriarchy without necessarily committing to action or redistributive justice. Thus, self-reflexivity as a moment of admission can be interpreted as a sign of transcendence (e.g., as white male, I recognise forms of racism). However, it can in fact be a non-performative or an "unhappy performative": "the conditions are not in place that would allow such 'saying' to do what it 'says'" (Ahmed 2006, 5). Such reflexivity allows individuals to disavow responsibility for the collective, and maintain forms of unseeing. Moreover, through the reflexive process, we risk authorising the "expert" or superior academic and the subordination of the researched (Wray-Bliss 2003). In reconceptualising and redeploying reflexivity, it must not be a "personal attribute, character disposition or acquired skill, but rather, ethically, it is an existential struggle that is at the heart of any practice that would involve generating knowledge about other people" (Rhodes and Carlsen 2018, 1304). Here, I join critical feminist postcolonial researchers (e.g., Back and Puwar 2012; Smith 1999), in reflecting and deliberating on the politics and ethics of craft research practices (Bell and Willmott 2020).

I turn to Judith Butler's work (2016, 2020) on vulnerability to explore what it may offer in the way of an ethico-political imperative to examine social suffering, including whose suffering is deemed grievable and who is denied grievability. I draw on my research examining the community-based palliative care movement in the state of Kerala, India. Over ten years, and with multiple collaborators, I examined how communities organise to provide palliative care for terminally and chronically ill patients and their families, which improves their quality of life. Drawing on specific moments from this research process that have impelled me to stop, step back, start again, retrace, I reflect on how one may inhabit spaces of trauma, illness, loss, and suffering. The objective here is not to develop a new set of methodological prescriptions or "best practice" recommendations. Rather, I trace the messy, entangled, embodied work of researching social suffering. These reflections have been shaped by dialogues, debates, and writing with academic co-authors, but also generously by multiple stakeholders in the palliative care field.

Vulnerability as praxis

Vulnerability can be understood as an existential condition, wherein we are all subject to accident, catastrophe, attack, or illness. It is also a socially, politically, and economically produced and managed condition associated with underlying disproportionate suffering and precarity. Vulnerability is then a "relation to a field of objects, forces, and passions that impinge or affect us in some way" (Butler 2016, 25). Butler makes a crucial distinction in her conceptualisation of vulnerability and precarity, noting that there are differential precarities, and it is important to distinguish between the precarities of those with differential access to housing, water, healthcare, and so on. Precarious lives are constituted by these vulnerabilities to the actions of others through our relations of interdependence.

As a shared ontological condition, vulnerability acknowledges that the human body exposes us to others' actions and makes us susceptible to a spectrum of responses, including violence and abuse, care and generosity. This common dependency and interdependency form the basis of our vulnerability – a way of being related to "what is not me". Attending to ontological vulnerability and relations between vulnerability and interdependence becomes a provocation for an ethics and a politics (Murphy 2011) that compels us to revisit and reframe our methodological premises. This ontological vulnerability of a human body opens us to moral and ethical obligations to the claims of others, to alleviate violence and suffering (Murphy 2011). Butler (2020, 88) writes,

So, the situation of many populations who are increasingly subject to unlivable precarity raises for us the question of global obligations. If we ask why any of us should care about those who suffer at a distance from us, the answer is not to be found in paternalistic justifications but in the fact that we inhabit the world together in relations of interdependency. Our fates are, as it were, given over to one another.

This vulnerability then for Butler is an ethico-political position, constitutive of an egalitarian imaginary that recognises the interdependency of lives. Vulnerability as praxis calls into question why certain forms of suffering are deemed suffering and how the recognition of that suffering in legal and human rights discourse is imbricated in the possessive individualism characteristic of capitalism (Butler and Athanasiou 2013).

Importantly, for Butler, vulnerability is not about embracing feelings, revealing our individual fault lines, a mode of being more authentic, or a subjective disposition. Rather, vulnerability is a mode of being simultaneously exposed and agentic. Indeed, vulnerability can be an "incipient and enduring moment of resistance" and is not in opposition to resistance or power (Butler 2016, 25). Vulnerability is not weakness, passivity, or victimisation but is the condition of the very possibility of resistance. Resistance draws from vulnerability; vulnerability is part of the very meaning or action of resistance.

If our mode of reflexivity is "stylized and maintained as a social and ethical practice" (Butler 2009, 113), then vulnerability is a certain performative production within these established conventions and cannot be in and of itself a departure from conventions. In other words, there is nothing in the acknowledgment of vulnerability that is inherently ethically prescriptive: the acknowledgment of vulnerability can inspire generosity and love as much as it can inspire violence and dispossession (Murphy 2011). For vulnerability to be transformative praxis, as theory and action directed at transforming the world around us, it must inform our self-examination, the account of oneself. Vulnerability as praxis, asks us, "how to become dispossessed of the sovereign self and enter into forms of collectivity that oppose forms of dispossession that systematically jettison populations from modes of collective belonging and justice" (Butler and Athanasiou 2013, xi). In the following sections, I draw on my research on Kerala's community-based palliative care movement to think through vulnerability as praxis.

Kerala's community-based palliative care movement

The story of the community-based palliative care movement in Kerala is an account of how ordinary people – schoolteachers, farmers, small traders, bus and auto-rickshaw drivers – forged solidarities with nurses and doctors to transform the boundaries of palliative care (Vijay, Monin and Kulkarni 2020). With its genesis in a doctor-driven clinic at the state-funded Kozhikode Medical College in 1993, the movement expanded to hundreds of decentralised community organisations across Kerala. Community organisations, comprised of volunteers, doctors, and nurses, provide "total" home-based care – i.e., financial, social, emotional, bereavement support – free of cost to terminally and chronically ill patients. Through these community

organisations, palliative care coverage in Kerala crossed 60% of the patient population, in comparison with the rest of India, where palliative care coverage remains
at less than 2% (Vijay and Monin 2018). Lobbying and advocacy by community
organisations translated into a state policy on healthcare, reflecting a profound case
of bottom-up claim-making. The state government today is the biggest provider of
palliative care in expenditure and number of organisations, with primary health centres and district hospitals running palliative care units. The policy mandates the state
public health system to collaborate with local community organisations for effective
care delivery.

Central to this movement are people's solidarities across differences – of class,
gender, religion, language, and political commitments – creating bottom-up
claims on the state (Vijay, Monin and Kulkarni 2020). In the floods that ravaged
Kerala in 2018 and the ongoing COVID-19 crisis, the state mobilised palliative
care networks (Praveen 2020). In this context, understanding a movement that
has communities working symbiotically with state infrastructure and deepening
democratic processes through bottom-up claim-making is essential to understand
how to resist and imagine alternatives to neoliberal forces of privatised healthcare.

My engagement with this movement began in December 2009 as a doctoral
student interested in alternative forms of organising, and specifically community
organisation. My dissertation traced the emergence and evolution of the frames
and the organisational field of community palliative care. Thereafter, various questions have propelled this journey, some arising from academic intrigue, others in
dialogue with palliative care providers. How did subordinate and dominant actors
forge solidarities to construct an alternative organisational form that alleviated
social suffering? What are the structural conditions that enabled the community
to form in Kerala? How do we understand multiple conventions and orders of
worth inhered in caregiving? How can we translate a community form to another
institutional context?

Between December 2009 and July 2019, I conducted multiple field visits, interviewing over 140 stakeholders. An important part of this account is over 1200
secondary data sources, constructed over the years. Secondary sources include
organisational documents, training manuals, pamphlets, handbooks, brochures,
newsletters, patients' and volunteers' poems and craftwork, newspaper articles, and
technical reports in medical journals and institutional outlets (e.g., World Health
Organization reports). Secondary sources have been crucial in piecing together the
movement retrospectively, tracing the founding of a clinic, policy changes, shifts in
physicians' perspectives over the years, and so on.

I participated in over 40 home care visits (each visit typically covers 4–5 patients
in a day), attended numerous monthly and home care review meetings with community organisations, and attended a communication training event for volunteers.
Navigating an insider-outsider role within the field (Sherif 2001), I participated in
national palliative care conferences as a presenter as well as panelist, provided qualitative methods training to palliative care professionals, and conducted research talks.
In 2014, Suresh Kumar, one of the pioneering palliative care physicians in Kerala,

informed me about his organisation's involvement in a translation effort at West Bengal (where my workplace is located, albeit in another district). He asked, "Do you want to document this?" What followed was an exciting phase of studying the seeding of a community palliative care initiative in a vastly different context from Kerala. This allowed me to seek comparisons from physicians and trace the disparities in healthcare contexts within India (Vijay, Zaman and Clark 2018). Around this juncture, I began collaborating with the End of Life Studies group at the University of Glasgow on their "Global interventions in the End-of-Life" project and, more recently, on developing a conceptual model for evaluation of Kerala's community palliative care. This evaluation exercise is explicitly designed to interrogate and mitigate cultural imperialism in evaluation exercises and to develop a model based on stakeholder criteria – not just imported global benchmarks of care delivery. These collaborations have breached disciplinary boundaries by working with critical management scholars, sociologists, anthropologists, and physicians. In these ways, through its ongoing dialogue with social theory, palliative care providers, and recipients, this body of work seeks to contribute to a public value approach to social theory (Delbridge 2014).

In the following sections, I explore how vulnerability enables adoption of three modes of research praxis: (1) *vulnerability as susceptibility*, which allows openness to silence and challenges epistemic certitude; (2) *vulnerability as collective care*, which acknowledges the role of time and generosities; and (3) *vulnerability as learning to be affected by difference*, where one learns from wounding and unsettled habitations that arise over the course of fieldwork. I draw on examples and anecdotes from fieldwork to exemplify these three modes.

Vulnerability as susceptibility

Vulnerability as susceptibility recognises that we are all "unknowing and exposed to what may happen, and our not knowing is a sign that we do not, cannot control all the conditions that constitute our lives" (Butler 2015, 21). Vulnerability, thus understood, informs our methodological practice by creating space for "not knowing" – the non-consensual, the discontinuities, surprises, and silences. Vulnerable praxis acknowledges the dispossession of intelligibility, the moments of unknowingness and unknowability in our research encounters. Struggling with the incoherence, opacities, and contingencies of our self-examination then constitutes the basis of an ethico-political stance (Butler 2009).

As a part of an exercise in conceptualising a framework for evaluating Kerala's community palliative services, I need to understand patients' experiences of care provision and what they value. Prakash, a trainer for volunteers, and Vishnu and Wafa, both community volunteers, are assisting with fieldwork. They suggest interviewing Shazia and Naz, who had lost their three-year-old daughter to cancer a few years ago. In February 2019, we meet Shazia and Naz at their home in Kozhikode city. Their older son is out for school, and their two-year old daughter plays around us during the interview. The volunteers know the

7

couple well. Wafa informs them of her upcoming marriage. Prakash speaks with them about their participation at a palliative care event he had coordinated the previous month. Shazia asks me about my work. She informs me that she is a kindergarten teacher and volunteered with terminally-ill children admitted at the medical college several years before they had their second child.

SHAZIA: . . . after a few years, we planned a second child. We had a baby and then when she was 1.5 years, she was diagnosed. And then . . .
[pause. tears.]
Sorry . . . [pause]
We had our treatments done from RCC [Regional Cancer Centre] in Trivandrum.

The child was treated at the Regional Cancer Centre for eight months. After 1.5 months, the cancer recurred.

SHAZIA: We went through all the hardships. We know how those six months passed. [pause]. I have seen children getting around 30 shots a day, so that is terrible. We did not want to go back. She could not sleep. She did not sleep at all. She was disturbed, she used to cry in pain, but then we did not even think it is, never had a feeling that it would recur in 1.5 months. We were expecting it after a year maybe, but definitely not so early. So that was very . . . [voice trails off]. Then we went to palliative care.

Naz moves in and out of the room, never sitting for more than a few minutes. He fixes all of us a "nutritional drink" of milk, bananas, and *avil* (flattened rice), and moves out of the room again. About 45 minutes into our conversation with Shazia, Naz finally sits down. Prakash shares that Naz had attended a death café at the community clinic last month. I converse with Naz about the death café – what did they do, what did they talk about, who was there? Naz said, "People have fear when they speak about death, but when they understand other people's pain, then they come out with their stories." We speak at length about the structure of the death café and others' stories. Naz finally slips into telling his own story and doesn't stop for another hour:

[For the child] there was a tube to pass urine; motion was not okay. Surgery was just done ten days back. The stitch from the surgery was yet not healed. It was a bad condition . . .

Through over two hours of our conversation, the child is never named. We speak in their living room, the corridor, the veranda with dense foliage tumbling from the terrace, the kitchen. The parents talk about the disease, the intimidating, foreboding gates of the Regional Cancer Centre (Thiruvananthapuram), the long walks from the gate to the cancer wards, the endless hours in the waiting room, treatment, recurrence, pain, treatment, relapse, their decision not to treat, palliative care, and her last week of life. They say the loss has brought them closer. Their accounts fuse, then depart. Both Shazia and Naz now volunteer in varying capacities at the local palliative clinic. Indeed, Shazia accompanied us to the next patient we interviewed, someone she knew but had not met for a long time. We had come by an autorickshaw, so Naz decided to drop us in his car and invited us to meet again at the end of the day for dinner at a hotel in the city. And so this caravan moved, picking up and dropping off people along the way.

Openness to silence

I am entering the parents' space of suffering, death, and grief as a stranger. While I seek to understand the organisational aspects of care provision, this interview is intertwined with a collateral excavation of memories of their loss. Bourdieu writes (1999, 623), "The intrusion of the analyst is as difficult as it is necessary." How must I inhabit this space, acknowledging the intrusion of my presence here? A condition of vulnerability as susceptibility allows for an openness or surrender to unknowingness in the face of such intractable grief. What do I do? When must I interject? When must I not? What should I be asking? What should I not ask? Far from prescribing silence as an instrumental approach to research encounters, I am acknowledging the silence that arises in those moments of unknowingness, of not knowing what to say, or how to respond to the suffering of terminal cancer and loss of one's child. Patients' trauma, pain, and suffering, although the object of several profound inquiries, are difficult to translate and beyond the boundaries of disciplinary knowledge productions (Gunaratnam 2012). Of not needing to fill this moment with "knowing" or the "skill" of what to say, or what to do with one's hands and eyes, one's very presence. In this encounter with pain, illness, and social suffering, one encounters the limits of knowingness.

Acknowledging unknowingness through silence is not a naïve surrender or a convenient means of escape from complicated, emotionally intense situations. This would be another kind of silence, a more violent one, an act of force through the unwillingness to engage with social suffering, an indication of complicity in the societal mystification of social suffering. As Toni Morrison (2019, viii) writes, in responding to situations of uncertainty and chaos, we may respond with stillness, which may be "passivity and dumbfoundedness," "paralytic fear"; however, "it can also be art." Indeed, palliative care clinicians learn to engage with silence skilfully and as a contemplative practice in their conversations with patients (Back, Bauer-Wu, Rushton and Halifax 2009). Allowing the steadfastness of silence to do its job, acknowledging that "antiphony of language and silence" (Das 1997, 68), can make understanding and explanation possible.

Challenging epistemic certitude

In contrast to a position of epistemic certitude, vulnerability as praxis involves approaching a research site by acknowledging the limits of one's own knowledge and with an attitude of self-examination based on openness to the encounter. This modality is less about passive openness or examining the limits of one's thinking or prejudice. Rather, one's relation to the other as interdependent becomes the foundation of praxis. Rhodes and Carlsen (2018, 1297) write, "[v]ulnerability is hence a radical openness to the other that serves to put the knowledge and self of the researcher in constant question." How do I translate that grief into words? What words will carry the heaviness in Shazia's voice – heaviness flowing from a strength I cannot think through or imagine? What textuality surfaces the impenetrable pauses that contain within them memories of their child and her incurable pain? How does one respond to the grief of a stranger one has only just met but is so generously and intimately being shared? To even consider writing about this appears as a violation or revelation of an encounter based on a degree of trust (Bourdieu 1999).

Unknowingness continues to haunt vulnerable praxis in writing as well. How quickly, impelled by the imperative to write, or sheer arrogance of position, do I move towards epistemological certitude? Moving backward, reconnecting dots, moving forward from, I have entered a story, recrafting it through memory work. Vulnerable praxis then involves surrendering the "egotistical comforts of one's own epistemic authority" (Rhodes and Carlsen 2018, 1311). This condition of vulnerability counters the logic of mastery, numeration, and calculability (Butler and Athanasiou 2013) that characterises neoliberal structures of knowledge production.

Vulnerability as collective care

Vulnerability enables us to embrace a self that is passionate, as "dependent on environments and others who sustain and even motivate the life of the self itself" (Butler and Athanasiou 2013, 4). These care collectives, variously conceptualised as friendships in social justice activism (Banerjea, Dasgupta, Dasgupta and Grant 2018), affective communities (Gandhi 2006), dissident friendships (Chowdhury and Philipose 2016), or epistemic friendships (Nguyen, Nastasi, Mejia, Stanger and Madden 2016), are integral to social transformation – seeding generosity, empathy, love, affection, and emotional entanglements. These deeply affective connections are often obscured in the rational, deliberate exercise of reflexivity (Rhodes and Carlsen 2018) and the "rigor" and "truth-making" of scholarship (Chowdhury and Philipose 2016). A vulnerable praxis enables us to depart from individualistic academic practices (Höpfl 2000), reminding us of the collectivities we forge and how our own writing for justice, for change, is collectively imagined (Alexander and Mohanty 1997; see also Vijay, Gupta and Kaushiva 2020).

Here, the relations go beyond instrumental, transactional exchanges. This may mean slipping from formal interviews into hours of conversations outside on the porch, during bus rides, between home care visits, sharing tea and meals. The end of a workday may involve meeting a participant's significant other for coffee, stopping by their house to meet children, waiting to try out "Mandhi" an Arabic rice that is a local favorite served at the new restaurant that has opened on our field route, getting lost in streets between interviews to check out graffiti together, sharing grilled fish at the beach at night, talking about heartbreaks and long-distance marriage. Rich insight has come from those moments where formal tools of methodological practice have been suspended. During these moments when the researcher's role is in abeyance (albeit with the researcher's script always lurking), I learn how student volunteers spend their leisure time at palliative care organisations; some start dating and marry other volunteers. I am invited into their worlds to glimpse how spaces of care offer alternative imaginaries to youth that depart from the consumptive spaces of coffee shops, cinema theaters or catering at "rich people's parties" to earn money. This is not to romanticise spaces of care, imbricated as they are in worlds of publicity, fame, or recognition (Boltanski and Thévenot 2006), where selfies on Facebook with patients or other volunteers offer another neoliberal currency. Drawing on these experiences, we co-wrote about how multiple worlds – the civic, market, fame, industry, the inspired – co-exist in transformative service ecosystems. Significantly, these chance encounters and in-between conversations enabled us to glimpse moments of love, sacrifice, and of giving without expectations of return (Varman, Vijay and Skålén 2021).

Care and time

Friendships, entanglements, and relationships of intimacy in research take time – to learn to speak, think, talk, but also to just be there. This commitment to time interrupts the speed of the neoliberal process of writing papers, churning out more for productivity (see Kothiyal, Bell and Clarke 2018). Elsewhere, talking about the crisis in the humanities, Gayatri Spivak says (2016), "You cannot do moral metrics by knowledge management techniques. You have to cook the soul slow." During finite days of fieldwork, these spontaneous moments of affection and connection interrupt efficiency, speed, productivity, enumeration, and calculability, underpinning capitalist structures of the research process.

> One afternoon, back from a field visit, we are unwinding at the amphitheater at the Institute of Palliative Medicine, Kozhikode. I had a Nikon DSLR I had been using for filming a documentary. Student volunteers are hanging around. Friendly banter. Periodically one would disappear and take a patient in a wheelchair around the campus. Two young women ask if I can take their photo for a "DP" (display picture for social media), which turned into a photo op for everyone. They want me to share the photographs later. While we take pictures, one volunteer climbs a tree to take down an entangled banner. Community organisations talk about creating autonomy for volunteers. Participants tell me in interviews that they volunteer because they feel "ownership," it is their "responsibility." When I speak to the volunteer later, I learn that the task was not assigned to him, no one had even noticed the banner, but the event was over two months ago. In these idle moments, in the slowness of an afternoon not "doing" productive work, of not searching, probing, we stumble into "ownership" and "autonomy" as abstract, contested categories in community participation, enacted in fleeting, mundane everyday activities, as the volunteer climbs the tree and takes the banner down.

FIGURE 4.1 Student on the tree

Care and generosities

These care collectives, at times intimate, mostly loosely coupled, have crucially nourished and sustained me through the vagaries of the research process. I have struggled as an organisational theorist to make sense of the relevance of my research, and its policy and practice implications (cf. Delbridge 2014), partly estranged by the Global North centricity of organisation studies discourse. Such friendships and affective commitments can inspire work, generate new intellectual and political commitments that do not necessarily fit within silos of disciplinary boundaries, and can offer respite from the alienating aspects of imperialist knowledge production. These care collectives are infused with generosities.

These affective commitments, as with friendships, can involve sacrifices, a giving up, or a giving in. For instance, during a tightly scheduled field visit, primarily to meet with government functionaries at Trivandrum, I did not schedule meetings with other participants. Saif, a volunteer since 2006, and by then a good friend, squeezed time out of his work to meet for coffee in the evening. We shared updates on our work. I told him about an unusually tumultuous encounter (discussed in the next section) that had rattled me. Saif found it funny, and recounted numerous incidents that were absurd and hilarious in hindsight. I wanted to turn my recorder on because he was sharing gems from the early years. If I had asked, I knew he would have said, "Yeah, yeah, sure." But that single act of turning on the recorder would have instrumentalised, perhaps ripped through, the fabric of our conversation. So, it was going to be coffee and reveling in that moment with this friend. Without a recording, I would fail to recollect the thickness of Saif's account, his intonation, words in the vernacular. But these moments from the early years, revealing finer elements of community participation, elements of joviality and hilarity in a context of suffering otherwise shaded in disease and grief, are rarely revealed in formal interviews. These absurd anecdotes are hard to elicit in interviews unless one knows what to ask, and even then, may be elided. It is in these easy moments of two friends sharing a cup of coffee that such memories are excavated, unsolicited and unasked.

At times when neoliberal academic labour can overwhelm, the commitments of thousands of such caregivers impel me to put the next question to paper. How does one talk about empowering research methods without acknowledging the generosity of participants to my research? I am dependent on that sustenance, on that care, on that generosity. How do I factor in these affective commitments into methodological accounts that typically begin and end with the imperative to linearly and rationally "collect data"?

Such collectives recognise the difference and do not necessitate that we identify with one another. Moreover, one has to be careful that this care trope does not reinforce neoliberal imaginations of individualism, self-help, and self-care. Admittedly, these collectives do not necessarily subvert the status quo or radically challenge capitalist or patriarchal processes. But in a context rife with atomisation, instrumental, predatory relationships, and careerism, there is something about such care collectives that just slows down that capitalist process, perhaps a little bit. Not so fast. Not today. Because we care (see Figure 4.2).

FIGURE 4.2 Because I care installation

Vulnerability as learning to be affected by difference

For counterhegemonic politics, vulnerable praxis must necessarily attend to difference. Vulnerability is based on our interdependencies, "capacity to affect and be affected," implying the embodied affective character of subjectivity (Sabsay 2016, 279). The challenge is to consider this affective, embodied dimension of political subjectivity while accounting for hegemony. Vulnerability as *learning to be affected by difference* involves a constitutive openness, acknowledging that we are marked by our encounters with others, taking responsibility for others' histories and our own implication in these histories. With vulnerability as a relational perspective, Rhodes and Carlsen (2018, 1297) note, "the other is not considered another of me, capable of being captured in my imagination and acts of signification; it eschews a drive for that which would invoke classifying and comparing other people in one's own frameworks of knowledge." From this condition of vulnerability, one may avoid a denial of difference, a separation from the researcher and the person who is suffering. The objective then is not to erase difference but to learn from the wounding, to learn from being unsettled.

In the absence of critical reflection, this methodology can amplify the powers of dominant institutions by projecting vulnerability to the Other or by deeming vulnerable communities as without agency. Such a disavowal of vulnerability enables dominant actors to shore up power and rationalise the subjugation of those deemed vulnerable. This is typified in calls by humanitarian agencies to rescue or help vulnerable populations or in research programs around vulnerable communities, thereby reifying borders of injurability. Such a perspective is depoliticised and does not consider the causes of inequality and how we are all implicated in the differential production of vulnerability (Sabsay 2016, 279) through colonialist, casteist, or capitalist production processes.

One case that cast into sharp relief the question of difference within the movement, and between the researcher and the researched, was an encounter at one of the earliest community organisations in Kerala. I had visited this clinic in 2010 and returned in 2019 for the evaluation project. Vishnu, helping me with fieldwork, had arranged a meeting for us the next day.

Abdul *ikka*,[1] Samina and Raashid met us at the community organisation at 8:30 am on a Friday. I had not met them before, although Abdul *ikka* was often referenced by other interviewees. All three had been volunteering for nearly two decades. Samina had trained and become a palliative care nurse now. All morning, we reconstructed the history of this remarkable clinic, going back to the earliest days. Mid-morning, Samina said they were cooking a dried fish delicacy for us and that we must stay for lunch. Other volunteers returning from home care spoke with us. Closer to noon, we took a break. Samina pulled out photo albums of the early days (nearly twenty-two years ago), which I was delighted to browse through, thrilled to identify those I had interviewed. Immersed in these photos, I had seen a tall, bearded man enter the common area where we were huddled around a table. I smiled, acknowledging his presence. He seemed familiar, but I couldn't be sure. An introduction didn't seem on any one's minds, so I continued talking to Samina and Abdul *ikka*. At some point, the bearded man interrupted us and asked me who I was. I replied and went back to my interrupted conversation with Samina. Shortly thereafter, the bearded person raised his voice and asked whether we had permission to be at the community organisation. Although the question was strange because some of the oldest members were with me, we had spoken to Fasil before coming. Since it was a Friday, Fasil was busy at the mosque, but Abdul *ikka* would be there to meet me.

Samina told me he was Amjad. Amjad was an early volunteer. I had interviewed him at length in 2010 – he was a dedicated volunteer, champion grassroots organiser. Yet, nine years on, I didn't recognise him. In my memory, Amjad was lean, frail, and embodied his working-class background. The person in front of me was decisively well-built. By now, Amjad was raging, his voice ricocheting across the walls. He expressed his displeasure regarding research. I was confused. This was the first such encounter in all these years across community organisations. Plus, the oldest volunteers there had welcomed us. We had spent a warm morning together, as was usually the case in any community organisation. I apologised to him for not recognising him right away, even though I had interviewed him several years ago. I explained that I didn't expect him to remember me, but I knew of his role in the movement.

But now Amjad was talking about how some physicians had got awards, and urban community organisations had received media coverage for their work in palliative care. Amjad remarked, "Palliative care is of the community. How can one individual get an award?" (quoted from Varman, Vijay and Skålén 2021). Amjad raged about how he still held these physicians in high regard for past work. But such actions meant that they (him speaking for the community organisation) had to be careful with whom they spoke. Everyone else was silent around us. As he spoke, it was increasingly clear that his anger was directed at the more privileged – physicians and city-based volunteers. I doubt that Amjad desired the same recognition – to him, it was a violation of the basic tenets of community work. Here I was, my middle-class privileges and position as a researcher likely to have proximity to physicians' perspectives, yoked to his disdain or distrust for "experts." This distrust was perhaps compounded by my not recognising him, fuelling his anger.

Samina told me lunch was ready. By now, I wanted to leave. My chest and stomach felt knotted up from this encounter. But I also knew of Amjad's commitment and his exemplary role, which made me understand why he was furious. I had no appetite. But they had been preparing fish all morning just for us. I followed Samina to the kitchen and asked her, "after all this, how do I eat?". Samina brushed it off and told me, "Aye, don't take it seriously (*aye, kaaryam aakanda*). Amjad *ikka* is like that." A younger volunteer who had been watching followed us to the kitchen and said, "Don't mind Amjad *ikka*. He has a temper. You must finish the work you came here for. Have your lunch." I have no memory of the fish I ate. Dead silence. Then small talk and a superfluous thanks to the cook for the fish curry.

As I came out, Amjad had cooled down and was standing by the door. There was small talk. "Do you want tea or coffee?" I declined. We spoke cautiously. While I left on a reconciliatory note, the incident left me unsettled for days afterward. Vishnu, also a palliative care volunteer, was rattled: "How can a palliative care volunteer speak like this? We speak of compassion and care."

Wounding

Learning to be affected, to be wounded – tells us of that wounding. What of that wound tells us of difference? Of our being disturbed? What comes to the surface of the skin from that wound? Page (2017, 24) notes, "What is of particular interest methodologically here is the way the notion of puncturing might help to interrogate epistemological assumptions, and its emergence through attending to the details and particularities within a medium." Vulnerable praxis leaves open the possibilities of not knowing the other. Derrida (1978, 121) states, "This incomprehensibility, this rupture of logos is not the beginning of irrationalism but the wound or inspiration which opens speech and then makes possible every logos or every rationalism." This wounding cannot happen, without at the same time telling us why we are wounded, unsettled, disturbed.

It was only after speaking with others in the movement that I could make some sense of this singular event with Amjad. Subsequently, discussing this incident with other stakeholders became a portal to understanding tensions and rifts between organisations that participants rarely talked about in detail. I had got caught in the midst of an older cross-fire. One volunteer recounted an incident, noting that "when it comes to palliative care, you are talking about compassion, and when it comes to a fight between two religious groups, you were vehemently fighting and physically abusing the other guy. There is thinking in separate boxes." One of the physicians wryly remarked, "We should look at whether 20 years of palliative care has made better persons in Kerala."

But Amjad was central to my construction of the movement as a people's movement. And unsettled as I still am by his aggression, I agree with his point – the palliative care movement was of the people. This point was invariably occluded in media narratives and in dehistoricised accounts within the movement itself that valorised heroic individuals, typically physicians. What had pricked me was that inasmuch as I had endeavored to dispossess power asymmetries as a researcher, I was interpellated as an unwelcome professional, likely to extract as others had done before, and in doing so, undermine the ethos of a people's movement. However, my accounts had consistently maintained the primacy of collective action and the role of subaltern actors. Amjad obviously did not know this, and his locating me with other professionals was wounding. However, the incident was pivotal in reinforcing three points on difference: (1) "palliative care was of the people." This was a movement built bottom-up, (2) the manifest class differences between the socially subordinated actors from working-class backgrounds and the middle-class physicians (3) that movement solidarities are rife with differences – of political commitments, ideologies, patriarchy. How we organise through these differences remained worthy of attention.

Unsettled habitation

Encountering difference during the research process involves a corporeal experience of being unsettled. These encounters across differences claim a decentering of self. This can involve unsettled habitations, occupying positions with unease, being uncomfortable in the research process with difference, with one's own privileges. Habitation involves recognising how my own situatedness shapes praxis. And yet, as much as habitations are important, there is a need to radically repoliticise belongingness, to step away from capitalist, patriarchal, heteronormative, and ethnocentric legacies, to perform alternative modes and sites of belonging, different from property and self-ownership (Butler and Athanasiou 2013). Vulnerability can help transgress habitations to understand difference.

In hindsight, despite my corporeal resistance to being there, I am glad I stayed for the fish curry, although I have no memory of its taste. I stayed in gratitude to Samina, Abdul *ikka*, and Raashid. As Bourdieu (1999, 609) writes, such research encounters inhere "a total availability to the person being questioned, submission to the singularity of a particular life history – which can lead, by a kind of more or less controlled imitation, to adopting the interviewee's language, views, feelings, and thoughts." To work across difference without ignoring these differences is affective. It makes claims of us. Pointing to this affective position, Bourdieu (1999, 614, emphasis in original) writes that such encounters

> can be considered a sort of *spiritual* exercise that, through forgetfulness of self, aims at a true conversion of the way we look at other people in the ordinary circumstances of life. The welcoming disposition, which leads one to make the respondent's problems one's own, the capacity to take that person and understand them just as they are in their distinctive necessity, is a sort of *intellectual love.*

These embodied and emotionally complex claims carry through field encounters to our academic writing, where "we put ourselves at stake" (Kiriakos and Tienari 2018, 266).

Vulnerability as praxis enables us to grapple with this giving oneself up for others. "Unsettled habitation" (a term borrowed from Gopal (2005, 6) and developed in Vijay, Gupta and Kaushiva 2020) is an ethical disposition that impels us to interrogate our desires to produce knowledge from another life's experience of marginalisation and suffering. This mode behooves us to ask how we may narrate without enacting epistemic violence through the representation of subalterns.

Conclusion

Contemporary research processes are sustained within neoliberal structures that limit and foreclose radical praxis (Alexander and Mohanty 1997). Our studies of social suffering are implicated in and complicated by these hegemonic structures. Vulnerability as praxis in studies of social suffering can be potentially transformative insofar as our engagement is a dialectical process of reflection and action, motivated by a desire to transform, change, or overcome exploitation and humiliation. Accordingly, vulnerability is an important political subjectivity in this transformative knowledge production that disrupts violent ways of knowing in the hope of an emancipatory ethics for an alternate, reimagined future.

This writing is grounded in an awareness of the contemporary crisis of neoliberalism and reflects on how one may resist or interrupt capitalist processes. Contemporary epistemic practices emanate from and valorise a condition of strength and mastery as the achievement of invulnerability. Such practices of mastery reproduce structures of domination and undermine forms of susceptibility that yield solidarity and transformational alliances (cf. Butler 2016). Accordingly, if the researcher's subjectivity serves a form of power that signifies masculinism and imperialism, erasing difference and enacting mastery and domination, then that subjectivity must be challenged and undone. Vulnerability as praxis becomes crucial to fissuring the subject, serving as "central to a politics that challenges property and sovereignty in specific ways" (Butler and Athanasiou 2013, ix). Accordingly, vulnerability as an ethico-political stance that recognises time and generosities of care interrupts the commodification and appropriation of suffering for profiteering and market rationalities. With a vulnerable praxis, epistemologically, we shift from paternalistic positions, "uses" of suffering, or the "spectacle of suffering" to being thrown into suffering and embodying interdependencies, allowing for *openness to silence* while *challenging epistemic certitude*. With this susceptibility, we encounter the limits of knowing, interrupting a disposition of epistemic domination by the researcher, compelling us to question our theoretical assumptions, analytical and political frameworks about the world. In doing so, vulnerability opens up a mode of resistance to contest dominant representations and epistemologies.

Further, vulnerability as transformative praxis suggests the possibility of working with others through difference – to see the compassion, solidarity, generosity, and love that co-exist in spaces of illness, trauma, and suffering. This critical application of vulnerability as praxis would reimagine the researcher's responsibility and accountability. Thus, while reflexivity impels us to inscribe an admission of class, gender, or race positions, without necessarily demanding commitment to action or responsibility (cf. Ahmed 2006; Swan 2017), vulnerability moves beyond reflexivity centred on the self. Vulnerability gives up "the notion that the self is the ground and cause of its own experience" (Butler and Athanasiou 2013, 4). The condition of vulnerability allows for the *unsettled habitation* that occurs when I give an account of myself. As Butler explains (2009, 8): "when the I" seeks to give an account of itself, an account that must include the conditions of its emergence, it must, as a matter of necessity, become a social theorist. The reason for this is that the "I" has no story of its own that is also not the story of a relation — a set of relations — to a set of norms." Vulnerability as praxis contests this deliberate and bounded selfhood, instead, opening up a scene of address with the interdependent relations and norms that shape one's living. In this giving an account of oneself, one takes responsibility, opening up the possibility of bearing witness to alternatives as a critical project to sustain and reinvigorate the politics of social justice. Challenging, wounding, unsettling, this giving an account of one-self in vulnerability can reshape the contours of intellectual and political practice. Here again, I turn to Butler (2009). Drawing on this condition of vulnerability in giving this account, one can hope that one has not been irresponsible, and even if one has, one will surely be forgiven.

Acknowledgments

I am grateful to various co-authors who have been part of this research. Numerous study participants have extended enormous help across the years. Aparna, Wafa, Vishnu, and Prakash were integral to the most recent round of fieldwork. I am thankful for their generosity and support. Thank you to Emma Bell for the incisive editorial comments that helped shape this paper.

Note

1 Ikka is an honorific for elder male, typically used among the Muslim communities in North Kerala.

References

Ahmed, S. (2006) The Nonperformativity of Antiracism. *Meridians*, 7(1), 104–126.
Alexander, J. and Mohanty, C. T. (1997) Introduction: Genealogies, Legacies, Movements. In *Feminist Genealogies, Colonial Legacies, Democratic Futures*, edited by J. Alexander and C. T. Mohanty, xiii–xlii. New York: Routledge.
Back, A. L., Bauer-Wu, S. M., Rushton, C. H. and Halifax, J. (2009) Compassionate Silence in The Patient – Clinician Encounter: A Contemplative Approach. *Journal of Palliative Medicine*, 12(12), 1113–1117.
Back, L. (2012) Live Sociology: Social Research and Its Futures. *The Sociological Review*, 60, 18–39.
Back, L. and Puwar, N. (2012) A Manifesto for Live Methods: Provocations and Capacities. *The Sociological Review*, 60, 6–17.
Banerjea, N., Dasgupta, D., Dasgupta, R. K. and Grant, J. M., eds. (2018) *Friendship as Social Justice Activism: Critical Solidarities in a Global Perspective*. Kolkata: Seagull Books.
Bell, E. and Willmott, H. (2020) Ethics, Politics and Embodied Imagination in Crafting Scientific Knowledge. *Human Relations*, 73(10), 1366–1387.
Boltanski, L. (1999) *Distant Suffering: Morality, Media and Politics*. Cambridge: Cambridge University Press.
Boltanski, L. and Thévenot, L. (2006) *On Justification: Economies of Worth*. Princeton: Princeton University Press.

Bourdieu, P. (1999) *The Weight of The World: Social Suffering in Contemporary Society*. Stanford: Stanford University Press

Butler, J. (2009) *Giving an Account of Oneself*. New York: Fordham University Press.

Butler, J. (2015) *Notes Toward a Performative Theory of Assembly*. Cambridge, MA: Harvard University Press.

Butler, J. (2016) Rethinking Vulnerability and Resistance. In *Vulnerability in Resistance*, edited by J. Butler, Z. Gambetti and L. Sabsay. Durham, NC: Duke University Press.

Butler, J. (2020) *The Force of Nonviolence: The Ethical in the Political*. New York: Verso Books.

Butler, J. and Athanasiou, A. (2013) *Dispossession: The Performative in the Political*. New York: John Wiley & Sons.

Chatterjee, S. (2020) A Suitable Woman: The Coming-of-age of the 'Third World Woman' at the Bottom of the Pyramid: A Critical Engagement. *Human Relations*, 73(3), 378–400.

Chowdhury, E. and Philipose, L., eds. (2016) *Dissident Friendships: Feminism, Imperialism, and Transnational Solidarity*. Champaign, IL: University of Illinois Press.

Das, V. (1997) Language and Body: Transactions in the Construction of Pain. In *Social Suffering*, edited by A. Kleinman, V. Das, and M. Lock, 67–92. Berkeley: University of California Press.

Delbridge, R. (2014) Promising Futures: CMS, Post-disciplinarity, and the New Public Social Science. *Journal of Management Studies*, 51(1), 95–117.

Derrida, J. (1978) *Writing and Difference*. Chicago: University of Chicago Press.

Gandhi, L. (2006) *Affective Communities: Anticolonial Thought, Fin-De-Siècle Radicalism, and The Politics of Friendship*. Durham, NC: Duke University Press.

Gopal, P. (2005) *Literary Radicalism in India: Gender, Nation and the Transition to Independence*. London: Routledge.

Gunaratnam, Y. (2012) Learning to be Affected: Social Suffering and Total Pain at Life's Borders. *The Sociological Review*, 60, 108–123.

Höpfl, H. (2000) The Suffering Mother and the Miserable Son: Organizing Women and Organizing Women's Writing. *Gender, Work & Organization*, 7(2), 98–105.

Khare, A. and Varman, R. (2017) Subalterns, Empowerment and the Failed Imagination of Markets. *Journal of Marketing Management*, 33(17–18), 1593–1602.

Kiriakos, C. M. and Tienari, J. (2018) Academic Writing as Love. *Management Learning*, 49(3), 263–277.

Kleinman, A., Das, V. and Lock, M. eds. (1997) *Social Suffering*. Berkeley: University of California press.

Kleinman, A. and Kleinman, J. (1997) The appeal of experience; the dismay of images: Cultural appropriations of suffering in our times. In *Social Suffering*, edited by A. Kleinman, V. Das, and M. Lock, 1–24. Berkeley: University of California Press.

Kothiyal, N., Bell, E. and Clarke, C. (2018). Moving Beyond Mimicry: Developing Hybrid Spaces in Indian Business Schools. *Academy of Management Learning & Education*, 17(2), 137–154.

Krumer-Nevo, M. and Sidi, M. (2012) Writing Against Othering. *Qualitative inquiry*, 18(4), 299–309.

Mohanty, C. T. (1984) Under Western Eyes: Feminist Scholarship and Colonial Discourses. *Boundary*, 2, 333–358.

Morrison, T. (2019) *Mouth Full of Blood: Essays, Speeches, Meditations*. Random House.

Murphy, A. V. (2011) Corporeal Vulnerability and the New Humanism. *Hypatia*, 26(3), 575–590.

Nguyen, N., Nastasi, A., Mejia, A. Stanger, A. and Madden, M. (2016) Epistemic Friendships: Collective Knowledges and Feminist Praxis. In *Dissident Friendships: Feminism, Imperialism, and Transnational Solidarity*, edited by E. Chowdhury and L. Philipose, 11–42. Champaign, IL: University of Illinois Press.

Page, T. (2017) Vulnerable Writing as a Feminist Methodological Practice. *Feminist Review*, 115(1), 13–29.

Praveen, S. (2020) A Soothing Touch at the Time of the Pandemic. *The Hindu*. www.thehindu.com/news/national/kerala/a-soothing-touch-in-the-time-of-pandemic/article31409439.ece

Rhodes, C. and Carlsen, A. (2018) The Teaching of the Other: Ethical Vulnerability and Generous Reciprocity in the Research Process. *Human Relations*, 71(10), 1295–1318.

Sabsay, L. (2016) Permeable Bodies: Vulnerability, Affective Powers, Hegemony. In *Vulnerability in Resistance*, edited by J. Butler, Z. Gambetti, and L. Sabsay. Durham, NC: Duke University Press.

Sherif, B. (2001) The Ambiguity of Boundaries in the Fieldwork Experience: Establishing Rapport and Negotiating Insider/Outsider Status. *Qualitative Inquiry*, 7(4), 436–447.

Smith, L. T. (1999) *Decolonizing Methodologies: Research and Indigenous Peoples*. London: Zed Books.

Spivak, G. (2016) *Critical Intimacy: An Interview with Gayatri Spivak*. An Interview by Steve Paulson. Accessed on 27th April 2020 from https://lareviewofbooks.org/article/critical-intimacy-interview-gayatri-chakravorty-spivak/

Swan, E. (2017) What Are White People to Do? Listening, Challenging Ignorance, Generous Encounters and the 'Not Yet' as Diversity Research Praxis. *Gender, Work & Organization*, 24(5), 547–563.

Varman, R., Vijay, D. and Skålén, P. (2021). The Conflicting Conventions of Care: Transformative Service as Justice and Agape. *Journal of Service Research*, doi:10.1177/10946705211018503.

Vijay, D., Gupta, S. and Kaushiva, P. (2020) With the Margins: Writing Subaltern Resistance and Social Transformation. *Gender, Work & Organization*, 28(2), 481–496.

Vijay, D. and Monin, P. (2018) Poisedness for Social Innovation: The Genesis and Propagation of Community-based Palliative Care in Kerala (India). *M@n@gement*, 21(4), 1329–1356.

Vijay, D., Monin, P. and Kulkarni, M. (2020) Strangers at the Bedside: Subaltern Solidarities and New Form Institutionalization. In *80th Annual Meeting of the Academy of Management Proceedings*, edited by Guclu Atinc. Online ISSN: 2151–6561.

Vijay, D. and Varman, R. (2018) Introduction: Undoing Boundaries. In *Alternative Organisations in India: Undoing Boundaries*, edited by D. Vijay and R. Varman, 1–25. New Delhi: Cambridge University Press.

Vijay, D., Zaman, S. and Clark, D. (2018) Translation of a Community Palliative Care Intervention: Experience from West Bengal, India. *Wellcome Open Research*, 3, 66.

Wray-Bliss, E. (2003) Research Subjects/research Subjections: Exploring the Ethics and Politics of Critical Research. *Organization*, 10(2), 307–325.

5

DRAWING ONE'S LIFEWORLD

A methodological technique for researching bullied child workers

*Ernesto Noronha, Premilla D'Cruz,
Saikat Chakraborty and
Muneeb Ul Lateef Banday*

Introduction

Though in today's ocular-centric world visual research methods are arguably the most powerful and versatile form of communication, they remain under-explored, under-theorised and underutilised by social scientists (Bell and Davison 2013Bell and Davison 2013; Guillemin and Drew 2010; Rees 2018; Ward and Shortt 2019). This underutilisation is surprising not only because of the ubiquity of visual images encountered in a wide variety of social contexts but also because visual skills and resources are taken-for-granted ways of being in the world (Harrison 2002). However, recently several academic disciplines have embraced the shift from the 'linguistic turn' to the 'pictorial turn'. This has been driven by the need for researchers to develop new forms of knowledge and understanding and more appropriate methods to counter the domination of language which has historically dominated social science research (Bell and Davison 2013; Rees 2018). Visual methods can help to overcome the logocentric tendencies of verbal research approaches and enable participants to express their views more directly with less interference from the researcher (Buckingham 2009). In fact, visual methods can assist in communicating complex nuances of lived experiences with simplicity (Bell and Davison 2013; Lenette and Boddy 2013). Using visual methods is important not simply because of the amount of data produced but also because of the quality of the data generated (White et al. 2010). Therefore, images need to be taken as legitimate objects of inquiry rather than mere adjuncts to linguistic meaning-making activities (Bell and Davison 2013). Besides this, visual methods may also be used to enhance data collection through rapport building and empathy, facilitating communication, enhancing expression of tacit knowledge, accessing difficult-to-reach participants, and encouraging reflection (Rees 2018).

DOI: 10.4324/9780429352492-5

Given the diversity of visually orientated disciplines such as anthropology, sociology, cultural studies, psychology, history and geography, it is not surprising that visual methods are also extremely heterogeneous and include both technology-mediated (e.g., photography and video) and non-technological tools (e.g., drawing, collage, playdough, lego and artwork) (Literat 2013; Rees 2018). A vast majority of recent studies have opted for technological tools such as photography and video, instead of non-technological tools. Thus, photovoice and digital storytelling have emerged as two popular approaches especially in work with children and youth (Literat 2013). Further, there are two distinct approaches to visual research techniques. The first uses pre-existing researcher-produced visual stimuli in an interview situation, and the second stimulates participants' field to produce their own imagery with respect to a certain issue (Pauwels 2015; Rees 2018). Visual methodologies have primarily involved the analysis of existing visual images or artefacts, or the study of images taken by the researcher at the study site. There has been limited research involving the study of images generated by participants as part of the research (Guillemin and Drew 2010). For instance, the use of images in modern social science originated in the field of visual anthropology, where researchers supplemented their narrative accounts of 'exotic' cultures through photographs. This was essentially an anti-participatory technique, placing the disempowered human subject under the gaze of the colonial researcher. Since then, photo-elicitation as a comparatively more participatory visual communication strategy emerged, where the primary objective is not documentation but the comprehension and analysis of an individual's intimate world. However, even under photo-elicitation, it is often the researcher that takes or selects the photographs, denying the participants the agency of producing the images themselves (Literat 2013).

Visuals as participatory

In contrast, a more participatory approach is enabled by allowing the members of the group or culture studied to produce their own images in response to a researcher-initiated assignment (Pauwels 2015). Such participatory approaches to knowledge construction necessitate a shift away from reifying traditional researcher-participant hierarchies that allow researchers to demonstrate expertise about participants with little or no participant input (Ross 2017). A participant-centred approach facilitates a process whereby researchers and participants work together in producing knowledge and meaning. Researchers gain knowledge by undertaking research with local people rather than on them (Oliveira 2016). Such methods involve the active involvement of both researchers and participants rather than passive involvement of participants in the research (Lenette and Boddy 2013). In fact, the influence of the researcher is restricted to a minimum, with only limited control over the way certain aspects are portrayed (Pauwels 2015). This also changes the researcher-participant relationships as participant-generated images emphasise participants as experts, shifting the balance of power away from researchers to the participants. Through this process, participant-generated visuals give voice to those who are typically silent (Rees 2018). The idea behind this is that research should also benefit those who are subjected to it and help to solve problems of communities without researchers thinking primarily about their own professional gains (Pauwels 2015). Thus, developing collaborative researcher-participant relationship is about empowering marginalised groups to transform their lives (D'Cruz et al. 2021; Oliveira 2016; Rees 2018).

Furthermore, important questions arise about who is interpreting images and for what purposes, and how the interpretation of images interplays with the interpretation of text (Rees 2018). In the case of participatory visual methods, both the participant and the researcher have a role to play in the analysis of the image. As the person who produces the image, the participant's interpretation of their own image is most significant. Therefore, the participant plays a reflexive role in both generating and interpreting their image. However, in terms of overall interpretation and analysis, it is the researcher who is the key (Guillemin and Drew 2010). Thus, participatory techniques provide opportunities to combine the 'emic' with the dominant 'etic' view as giving 'voice' to people is inadequate, the data needs to be analysed and framed within disciplinary methods and theory. Similarly, the absence of participants' views results in incomplete depictions and interpretations by researchers despite theoretical and methodological sophistication. In short, the expertise of researchers and participants needs to be skilfully combined to produce a more nuanced and situated understanding of the lived experiences of participants (Pauwels 2015).

This is enabled when participatory visual methods are combined with text-based methods. This combination can enable richer, more complex, deeper and sometimes hidden data rather than overly simplistic conceptualisations and misrepresentations that could occur from the stand-alone use of research tools (Oliveira 2016; Rees 2018). Further, the interview using visual stimuli can offer a wide range of relevant information about how respondents perceive their world. Combining these two techniques generates numerous opportunities for both knowledge production and social action (Buckingham 2009; Pauwels 2015). Through the process of visual conceptualisation, and the reflective discussion of these images in the context of their production, participants are also given an opportunity to voice their inner stories, as well as an active and empowering stake in the research study (Literat 2013). Finally, there are benefits in dissemination, as visual projects support the production of participant-generated and participant-selected material to be exhibited, distributed and shared with public audiences (Oliveira 2016). However, the challenge is to reach a text-image balance in analysing and reporting findings combining the 'emic' with the dominant 'etic' remains (Lenette and Boddy 2013).

In this chapter, we explain the use of drawings generated in collaboration with children working on cottonseed farms. However, before we describe our experience of the method, a brief note on drawings as a technique is relevant. Later, we conclude with a discussion on the use of participatory visual methods.

Drawings as participatory

Drawing stands in stark contrast to the manipulatable, manageable, calculable and modellable quantitative paradigm (Ward and Shortt 2019). They are visual products and at the same time produce meaning about how people see the world in both its simplicity and its complexities (Guillemin 2004). The act of drawing allows participants to feel more willing and able to share conscious as well as unconscious thoughts parts of their subjectivity in a way that words find hard to express (Broussine 2008). In this sense, drawings facilitate participants' expression of rich, deep discussions and emotions by enabling thinking, recall and storytelling about complex topics that are both unrehearsed and hard to reach by using methods such as interviews (Broussine 2008; Guillemin 2004; Ward and Shortt 2019). Further, unlike interviews or focus group where an instantaneous response is expected, drawings give additional time for reflection allowing participants an opportunity to craft a more complete depiction that encourages active conceptualisation and contemplation (Literat 2013).

Besides this, a major advantage of this methodology is its impressive versatility. It is a strategy that can be implemented easily and inexpensively and can be used where the language barrier between the researcher and the participant may seem like an insurmountable obstacle. Moreover, when compared to technologically mediated tools like cameras, drawing is comparatively more generative because participants draw a world into existence, unbounding their creative depiction of both physical and abstract realities (Literat 2013). It also enables the documentation of observational data when recording devices are not permissible and enhances the visual capabilities of researchers by communicating findings to end-users in powerful ways (Rees 2018). Clearly, because of its non-textual and highly accessible nature, participatory drawing promises to succeed where other methods could fail (Literat 2013). Furthermore, by allowing the participants to freely decide on the content and framing of their images, visual techniques can highlight both presences and absences, thus the significance lies in both the visible and the omitted (Literat 2013). Thus, participant-generated drawings broaden the scope of data access, and therefore open up the complexities of the phenomenon being researched (Guillemin and Drew 2010; Rees 2018; Ward and Shortt 2019). Besides this, participatory drawing is a comparatively more expressive, engaging and fun activity, which can turn the research study into an enjoyable experience for those involved and help maintain the participants' attention in situations where their enthusiasm or concentration levels are of concern (Literat 2013).

Furthermore, drawings are intricately bound up with power relations. This allows researchers to address a variety of power dynamics by giving voice to those who may not otherwise be heard. In this sense, the method takes participants seriously as knowers by fostering their participation and enabling their inclusion (Guillemin and Drew 2010; Ward and Shortt 2019). Nonetheless, there are concerns about interpretations of the findings. It is necessary to recognise that drawings do not 'speak for themselves' (White et al. 2010). Drawings produced as part of a collaborative inquiry method should not be treated as visual artefacts for objective deciphering; rather, it is a highly interpretative research method (Literat 2013; Ward and Shortt 2019). The quality of the dialogical engagement and discussion is as important as the drawing itself (Broussine 2008; Tay-Lim and Lim 2013). In this regard, it is worth remembering that participant-generated drawings resist a culturally neutral interpretation and are always a product of an individual's cultural background. Consequently, when the researcher belongs to a culture other than those they are researching, the temptation to over- or misinterpret can be particularly problematic (Literat 2013). Therefore, in order to handle the challenge of over or misinterpretation, drawn images must be interpreted by the participant or artist who produced the work and not by the researcher (Literat 2013; Ward and Shortt 2019). Besides this, allowing research participants to interpret their own drawings is not only more illuminating but also more ethical given that the visual evidence is a subjective product of the participants' own perceptions and lived realities (Literat 2013). This can be achieved through the complementary and symbiotic use of method triangulation such as interviews carried out either in a group setting or individually (Literat 2013; Ward and Shortt 2019).

The use of drawings within interview settings also enables the 'redistribution of power' between the researcher and the participant. Participants can shape their own agenda and direct the conversation according to their own feelings, reflections and responses to their drawings (Ward and Shortt 2019). This strategy of encouraging the participants to talk about their own drawings puts the participants in charge of the interpretation process and the researcher in the position of the listener. It also discourages the sense of a hierarchy between the researcher and the subject, and it is comparatively more horizontal and more ethical than alternative textual strategies (Literat 2013). In this sense, unstructured participant-produced drawing method is one of the most egalitarian forms of collaborative inquiry (Ward and Shortt 2019).

Not surprisingly, by facilitating communication and decentring the power relations of the research process, drawings open some of the ways in which children construct their identities (White et al. 2010). Drawings allow children to express themselves and communicate in ways that are not always wholly written or verbal (Guillemin 2004; White et al. 2010). By drawing the children take some control over the research process and prioritise issues that an adult researcher might see as irrelevant (White et al. 2010). In this sense, through drawings young children can enter the research process and be understood by researchers on their own terms (Tay-Lim and Lim 2013). In fact, the use of drawings in research with children assumes that they are active, competent and reflexive expert informants about their own lives (Tay-Lim and Lim 2013). Nonetheless, discussions with children who create these images are crucial to uncovering the complexity of these images (White et al. 2010). Therefore, both visual images and the verbal exchanges are central to the children's meaning-making process. It is necessary to discuss the drawing with the child, so that the child's meaning, and interpretation is represented, rather than those of the researcher (Literat 2013; Tay-Lim and Lim 2013). This facilitates the co-construction of children's ideas and reinforce their voices in research with adults and children as equal players (Tay-Lim and Lim 2013). Rather, children are in control and the presence of researchers is not required (Barker and Weller 2003; Bolton et al. 2001; D'Cruz et al. 2021).

There are two distinct ways in which drawings can be used. The 'draw-and-talk' method and the 'draw followed by talk' method. Some advocate the 'draw-and-talk' method, which they see as more promising than the 'draw followed by talk' method. The 'draw-and-talk method' records the journey of meaning-making right from the start of the drawing activity (Tay-Lim and Lim 2013). As the process of creating images is as important as the images themselves, the 'draw-and-talk method' involves asking children to draw a picture of what was important to them and to talk about their picture whilst they are drawing (Pauwels 2015; White et al. 2010). While the 'draw followed by talk' method, drawings are often used as part of a face-to-face interview wherein the drawings produced by participants are used by the researcher and participant to explore and examine together (Ward and Shortt 2019). However, in both approaches, participants are provided with a blank sheet of paper upon which they are asked to draw their thoughts, feelings and impressions in response to a particular question. They are often explicitly discouraged from labelling or using words but are reassured that as a part of the process they will have an opportunity to describe and discuss what they have drawn (Ward and Shortt 2019). Participants then discuss their images and meanings either with the researcher alone or with their peers (Ward and Shortt 2019). This allows researchers to develop theoretical insights from a 'grounded' perspective. An adult researcher takes on the role of co-constructor and allows children to demonstrate capabilities and understandings as active meaning-makers of their own circumstances. (Tay-Lim and Lim 2013; White et al. 2010).

Limits of visual methods

These methods are undoubtedly valuable, but they also raise some significant questions (Buckingham 2009). First, the major challenge is how to instruct the field and educate participants about producing visuals for a particular purpose without inadvertently transferring values and norms of the researcher to the participant (Pauwels 2015). Second, there are questions of individual skill. At the most basic level, drawing is a mode of communication with socially defined 'affordances' or potentialities. What counts as a 'good drawing' depends upon the conventions of representation and a given social context? Children or adults who perceive themselves to be 'no good' at drawing are likely to be constrained in particular ways. Further, there are significant differences in the drawing abilities of children of varying ages. Very young children generate simple scribbles but with age their dexterity and abilities to produce complex drawings increase (Tay-Lim and Lim 2013). Third, the idea that visual or 'creative' methods provide more accurate or authentic representations of individual 'beliefs' or 'attitudes' can be challenged. Data from visual methods may not be transparent evidence of inner mental processes and need to be analysed in terms of the context in which they were gathered

(Buckingham 2009). Fourth, several questions about empowerment and the potential for empowering research remain (Ross 2017). Claims that these methods are empowering for individuals in marginalised communities or that they are uniquely placed to give participants a 'voice' can be refuted (Buckingham 2009; D'Cruz et al. 2021; Oliveira 2016). Though participatory methods are more engaging and enjoyable, and alter the power relationships between researchers and researched, this hardly abolishes power completely (Buckingham 2009). For instance, research agendas are typically identified by funders, community-based organisations or by a research team rather than by the research participants themselves. Further, the identified 'problem areas' may not always focus on what participants feel are most pressing issues. Thus, the quest to produce and build knowledge about groups of people who occupy marginal spaces can be perceived, as an exercise of assumed power (Oliveira 2016). Besides this, most advocates fail to acknowledge the role of the researcher in both producing and presenting the material (Buckingham 2009). Fifth, participatory visual research approaches are time-consuming and intense endeavours for all involved. Asking participants to engage with their lived experiences may be invasive and raises issues of the appropriateness of such methods. Beyond appropriateness, participatory research often means that participants will share stories of trauma and personal accounts of pain and suffering, in addition to triggering negative memories and/or emotions for those listening, including researchers (Oliveira 2016). It also needs researchers to be prepared to handle emotions and feelings in empathetic, sophisticated and considerate ways (Guillemin and Drew 2010; Ward and Shortt 2019). Finally, drawings are inherently unfit for use with large groups of participants, and thus samples usually tend to be rather small, impacting the generalisability of the findings (Literat 2013).

With this understanding, we discuss the challenges and the issues we faced when using drawings with tribal child labour from the Bhil community working in cottonseed fields/farms.

Background about the community

The Bhils are one of the major Adivasi (first people) groups of western and central India. They consist of multiple sub-groups such as Bhilalas, Barelas and Naiks (Sinha 2017). These groups typically retain their own traditional and cultural practices and are marginalised from mainstream Hindu ritual praxis (Roche 2000). The Bhils inhabit the hilly regions of northeast Maharashtra, eastern Gujarat, southern Rajasthan and western Madhya Pradesh. This is denoted as the 'Bhil culture zone' because they once shared a common language, *Bhili*. The *Bhili* dialect of *Wagdi* is spoken in the southern Rajasthan region (where our participants came from) incorporates features of Gujarati, dialects such as Mewari and Malvi, as well as Hindi (Phillips 2012).

The common view is that the location of the Bhil community is an outcome of events during the feudal period when the Bhils were defeated by the Rajputs and pushed to the hilly forests forcing them to raid and indulge in dacoity.[1]

Consequently, Bhils earned considerable notoriety by making a living from robbery and looting resulting in others disrespecting and humiliating them (Sjoî`blom 1999). Therefore, they are often stereotyped by the urban upper-caste as uncultured, uncivilised and unclean 'hand-to-mouth' people who survived at subsistence level, without any thought of saving for the future. Besides this, they are also considered to be dangerous, liquor-drinking and wild people of the forest, armed with bows and arrows, highwaymen, thieves and dacoits (Mosse et al. 2005). Dwelling in the forest, Bhils made their livelihood from hunting, collecting forest produce, minor cultivation of grain or shifting cultivation and through selling wood, baskets, ropes and other small forest produce and providing protection to traders passing through their areas (Sjoî`blom 1999). These images of 'wild hill tribes' lent ideological justification for the British to rescue and tame them under their rule (Mosse et al. 2005).

Alternatively, others argue that the Bhil chieftains raided the plains to claim dues from peasant villages or to renegotiate Bhil claims in Rajput territories, which has been termed as 'shared sovereignty' over lost Bhil territory (Skaria 1998, 203). This changed with the advent of the British who crafted a 'new state-society relation' in which the Bhil became subordinated to a state (Nilsen 2015, 578). The British rule changed 'wildness' from a discourse of power into a discourse of marginality and the relationship between plains and hills from one of structured interdependence (in which raiding was a political act) to one of antagonism (in which raiding was a criminal practice contained by 'punitive expeditions'). In order to control and discipline the lawless Bhils, schools were opened, settled agriculture encouraged and the Bhil Corps[2] set up supported the settled agrarian lives of its members by advancing loans, discouraging alcohol and promoting education. This changing identity of Bhil communities was closely related to their loss of control over forest resources. First merchants and then the British took leases of the forest from Bhil chiefs, transforming the unruly forests into ordered high-value timber-producing reserve of teak. Besides this, forest was protected from Bhils and their hunting, gathering and shifting cultivation (Skaria 1997). Not surprisingly, over the last hundred years their livelihoods have transformed from forest-dwelling to settled agriculture (Sjoî`blom 1999). Further, to subordinate and rule, the British government persistently underscored the existence of logical, racial differences between India's caste and tribal communities. The ideas of primitiveness were equated with tribal communities, and that of civilisation with caste communities culminating in the creation of a list of Indian schedule tribes (STs) by the late nineteenth century and is still used by the independent Government of India to shape policy and practice towards these communities (Skaria 1997). Unfortunately, this state-society relation has persisted through time, with the Bhil finally being dispossessed and marginalised (Tandon 2019).

Not surprisingly, they mostly inhabit the undulating terrain and inferior land of the area, while other agriculturalists are found in the more fertile plains. Hence, the Bhils are forced to increasingly shift their livelihoods to wage labour (Sjoî`blom 1999). Further, as a result of land fragmentation and deforestation, agriculture has become precarious and Bhils have to supplement their resources through migration (Mosse et al. 2005). Not surprisingly, each year, thousands of young people from Adivasi communities (predominantly Bhil) from southern Rajasthan in western India migrate for seasonal work to several districts of Gujarat for labouring on cottonseed farms (McKinney 2014).

The study

The study was conducted on the behalf of Prayas Centre for Labour Research and Action (PCLRA) and was sponsored by Sudwind Institute, Austria. PCLRA and the *Dakshini* Rajasthan Mazdoor Union (DRMU), based on an earlier study conducted in 2007/08, have been spearheading the efforts. PCLRA, in a coalition with NCPCR, has engaged in sustained advocacy and campaigns to eliminate child labour found working on the cottonseed farms. Through the efforts of DRMU alongside NCPCR, mates, parents, schoolteachers and local communities are sensitised about the illegality of employing child labour. The subsequent visibility of the problem has led to some degree of decline in its incidence and to the rehabilitation of children trafficked to farms in Gujarat and Rajasthan and has also cautioned farmers about the issue (Banday et al. 2018). Nonetheless, the main objective of the current research was to conduct a follow-up investigation into the changes brought about in the working and living conditions of migrant Adivasi children from southern Rajasthan working on cottonseed farms in Gujarat.

However, before approaching children at the source areas, we made two short field visits to understand the production process of Bt^3 cottonseeds and the reason for employing children on the farms. From our several field visits we realised that organisers and farmers are aware that the employment of child labour is a matter of concern for the larger society and deftly dodged or denied even indirect questions on the topic. They denied using child labour and maintained that the children seen working on the farms were part of the farmer's family and were merely lending a helping hand beyond their school hours. When inspections and raids are carried out, the farmers manage to get the children to hide or run away from the farms during the duration of the check (Banday et al. 2018). Given the situation, it was decided to use drawings as a method to speak to children. With this background we planned to visit the villages to speak to migrant children.

Fieldwork

We began planning for our exercise a few days before our travel to the villages where the children lived. We bought some drawing paper, boxes of crayons, pencil and an eraser for each individual. On the day of travel, we started at 5.00 am, and it was nearly 10.00 am when we reached the first village. The villages were deep in the interior hills surrounded by the green and beautiful Aravalli ranges from all sides. We were helped by few local activists of PCLRA who belonged

to the Bhil community, and they took us to a house which was built at an eleva-
tion that oversaw the road. The inner side of the house was dark and dingy while
the exterior had a courtyard followed by an open space that was connected to
the road. We removed our shoes and seated ourselves on the floor which was
plastered with cow dung. The children mobilised by the activist had also gathered in the courtyard of this house waiting for us. On seeing us, the children
kept giggling and talking to each other as we explained the purpose of our visit.
They often spoke in meek voices, with some girls covering their mouths with a
duppatta or their hands and face held down. They were afraid, with one of the
activists telling us that the children thought that we were coming to take them
to the Bt cottonseed farms and were unwilling to come. Nonetheless, at the first
village, in order to build a rapport with the children, we offered them biscuits
and a few sweets before we started. However, this did not help in rapport building as the children were not interested in the treats but were keen to get along
with the task at hand.

We asked them to draw their experiences of work on a Bt cottonseed farm.
They were told not to use words and their artistic ability did not matter. As we distributed the drawing papers, pencils, erasers and crayons to the children at this first
village, we were confronted with several dilemmas. One lingering question was
how much direction we should give the children. After a bit of brainstorming, we
decided to be restrained in our instructions and gave the participants a broad remit
by asking them to draw about their workplace. We did not want to cue them too
much and instructed them to draw whatever came to their mind while working on
the Bt cottonseed farms.

All the participants agreed to draw but their level of enthusiasm varied. Some
looked around expressionless, while others got on with the task quickly. However,
we soon realised that giving a broad canvas was not helpful as they drew pictures
of cotton plants, plastic tags which were red in colour and the male flowers which
were strung together on a rod for cross-pollination, rather than the workplace relations in which we were interested. Some of them also wrote their names on top of
the drawing despite being told not to do so (see Figure 5.1). We soon realised that
our participants required us to be more direct and focus our instructions in line
with the research objectives. Consequently, we were more specific and told them
to focus on employer-employee relationships with an emphasis on bullying at work
and their life in general at the worksite.

At the first site, we decided to use the draw-and-talk technique. We tried to
speak to them as they drew but this did not help much. In response, many of them
only giggled. We soon realised that this intervention was not very helpful and
only harmed the process of them being themselves. Inadvertently, another distraction introduced at the first village was a stencil (that depicted a fish) which was a
part of the crayon box, and some of the children promptly began to use the same
without the fish being a part of their worksite (See Figures 5.1 and 5.2). Since the
crayon boxes were distributed before the researchers examining the contents, this

FIGURE 5.1 Child's drawing of Bt cotton seed plants, red tag, a fish and male flowers

distraction accidentally crept in. Following this, while visiting the other three villages, we saw to it that the stencil was removed from all the boxes.

In short, during this first round of drawings we learnt a few things:

1   We learned to be more directive in our instructions so that they focused on the problem at hand;

2   We realised that the draw-and-talk technique was a distraction and we should use the draw followed by talk technique in the next village;

3   In the next round, we decided to remove the stencil before handing over the crayons to them;
4   We also dropped the idea of offering them biscuits as they were more interested in drawing.

The impact of the reduced intervention was quite palpable as they were chuckling, talking and appeared to be enjoying the process. They also took the opportunity of less supervision to look at each other's drawings. In fact, we moved away from the location after the instructions were given, only to return time again in case the children required clarifications. In most cases they were comfortable and clear with our instructions. From a distance we could hear them speaking to each other and enjoying the process. We could also identify the most talkative child with whom we decided to do the first interview in the group. Overall, the entire process took three hours with children, who were taking at least 45 minutes to one hour to draw. Each group from the first three villages consisted of six to eight participants with the last group being more than ten.

The interviews

Unlike the photo-elicitation project which starts off with an interview with the participants, the main aim of the drawing was to act as a catalyst for the interview. Once all the children finished their drawings, we individually asked them to describe what they drew and why did they draw it. We had to confront several issues such as whether we do the interviews in a group or individually and who should we interview first. At the first village, interviews were held with individual participants, but later we realised that it was good to do it in a group, with each one discussing their drawings in the presence of others. This allowed them to open up to us and created a sense of safety where they could share freely. However, this approach was also fraught with the dangers of someone from the group commenting on the drawing and prompting the participant. This did happen a few times and the researchers had to instruct others not to prompt. But given that participants came from a marginalised community, it was better to continue in the group rather than individually. The group gave them the comfort of sharing their experience in the presence of others which was not very different. When children spoke in monosyllables, other participants were asked to aid in the process of our understanding. Even the quietest and youngest child were able to give significant accounts of their experiences at work. Drawings also acted as an evocative method of eliciting responses. For instance, the participant who created the drawing in Figure 5.2 was quite disinterested and drew an untidy picture of the cotton plant, a tag and a girl but was able to describe instances of bullying while working on the Bt cottonseed farm. If it was not for the use of drawings, it would not have been possible to do this research.

FIGURE 5.2 Child's drawing of Bt cotton seed plant, red tag, a girl child worker and a fish

Here is an excerpt of this interview:

Researcher: What is this?
Child: Kamal [flower].
Researcher: What is this?
Child: Girl.
Researcher: What is the girl doing? [No answer] How many girls were there? [No answer] Who is this other girl? Where is she from?

Child:	Kankariya.
Researcher:	What is this?
Child:	A tap.
Child:	How many kids were there?
Child:	40 workers (two mates). 10 boys and 10 girls. Mates stayed for 5–6 days and then left for home.
Researcher:	If you don't work what does Sheth do?
Child:	He beats us in such a case.
Researcher:	Have you been beaten? [No answer] What is this?
Child:	Room.
Researcher:	Are you alone here?
Child:	Others were also there but I have drawn only myself.
Researcher:	Why? Do you feel alone there?
Child:	There were others but I don't consider them. I don't have any connection with them.
Researcher:	Why so?
Child:	I did not see them as my own. I missed my parents.
Researcher:	When do you feel so?
Child:	All of these people were from another village. I was alone from my village. All of them used to scold me. So I felt alone. I did not feel good.
Researcher:	Why did they scold you?
Child:	There was too much work. I had been assigned a large portion of work. I was unable to handle that much of work. So Sheth used to scold me. Nobody stood up for me.
Researcher:	What did you do when you missed home?
Child:	I used to remember my parents and call my father. Some of the workers had phones, I called home from their phones. Father would ask [me] to come back when I shared that I don't feel good. But I could not go home all by myself.
Researcher:	Why did you not go back?
Child:	I feared how would I go back all alone by myself.
Researcher:	Why did you not like it there?
Child:	People there did not treat me well.
Researcher:	Did you make any friends there?
Child:	No.
Researcher:	When you were upset, did you talk to anyone there?
Child:	No.
Researcher:	When did you feel most upset?
Child:	I had fight with others a couple of times because I could not pollinate all the flowers.
Researcher:	Did you tell your parents that you feel alone when you returned?
Child:	Yes.
Researcher:	What did they say? Nothing.
Child:	They treat it as normal.

Further, the language Bhili, which to us sounded a mix of Hindi and Gujarati spoken by the participants, presented a major hurdle in communication. The use of drawings was the best way to overcome these handicaps. Besides this, since the children could not completely understand the language we spoke, we used the services of local activists (who also provided access) for translation, which helped immensely. For instance, one of the researchers used the term *doro*, which in Gujarati meant to draw; however, the translators suggested the term *banavo*, which means to make, be used because *doro* sounded like *dodo*, which means to run. The drawings gave the children some time to get used to us and then open up a bit during the interviews.

Following the interpretivist paradigm, the researchers accessed the range of constructions of meanings that children placed on their subjective experience as represented in their drawings. This unlocked the process through which their voice, feelings and recollections were heard (Broussine 2008). The questions asked pertained to not only what they had drawn but also about what had been rubbed off and if they wanted to share something which was missing in the drawing. We tried to make meaning of what they were drawing. In this sense, the drawings were used as data. The interviews were held so that meaning was made of the drawings and why they drew certain things. We asked: 'What does that show?', 'Why did you draw this?', 'What does that mean?' and 'Did you experience this?' This dialogue and the audio field notes were recorded at the end of each day. We wrote the names of the participants behind the drawings. When the group was less than ten, we waited for everyone to finish their drawing, but in the last site since the group was too large, we could not wait for everyone to finish their drawings as those who finished early were impatient to leave.

The pictures covered and clearly spelt out the context of their work. This created an opportunity for us to immerse into the life-worlds of child workers on the farms. The children drew the outlines of the farms with rows of cottonseed plant, their living quarters, wells, temporary washrooms and the tools used in pollination work. Some children drew figures of other children or adults sharing the living spaces, while some drew only themselves on the fields or in the living quarters, symbolising their feelings of loneliness and longing to go back home. However, they stayed back because of poverty faced by their families. Such stories not only evoked sympathy but also an appreciation of their grit. Besides this, the children also focused on the bullying they faced on the farm. For instance, in the drawing in Figure 5.3, the child draws the employer who scared the children and enforced control by brandishing a stick in his hand. This depiction of the employer (usually farm owner or sharecropper) with a stick featured in drawings of other children as well. Inquiring about this depiction elicited various stories of being beaten or witnessing another child being beaten or disciplined through fear. Nonetheless, children ran away to the cotton fields and hid themselves to protect themselves from violence.

In one case, the drawing stirred up painful memories. One boy wrote 'Diwali', which symbolised his bondedness. He missed celebrating Diwali with his family

FIGURE 5.3 Child's drawing of the employer who scared the children

members because the employer forced him to work against his will on a construction site by withholding his wages despite his several requests to be relieved. During the interview the child broke down, with the researcher promptly consoling the child (Figure 5.4).

FIGURE 5.4 Child's drawing of Diwali and a tagged Bt cotton seed plant

After the interviews in three villages we decided that we had reached saturation since participants repeated the same experiences, but we continued to collect more data since the activists had already fixed our meetings. To our surprise, we found a completely different group who had not migrated but worked on farms of their village or nearby villages. This provided a rich comparison with those who had migrated and told us a lot about their workplace and their everyday work. At the end of each round, we gave a slab of chocolate to all those who participated out of a sense of gratitude. Finally, in relation to ethics, there was no possibility of signing a consent form as the participants were poorly educated and insisting on such a document would vitiate the research process. It was enough for us to conduct the interviews under the gaze and help of local activists and elders of the community who had an interest in the Bt cotton industry. Finally, though the study was the property of Prayas, we had permission to publish the same for non-commercial purposes.

Discussion

The recent inclinations towards visual research methods are driven by the need for researchers to develop new forms of knowledge and understanding and more appropriate methods to counter the domination of language (Bell and Davison

2013; Rees 2018). Drawing is a unique way of collecting data because it conveys information in a way that words find hard or impossible to express (Broussine 2008). In fact, drawings provide insights into conscious as well as unconscious thoughts and feelings which participants interpret and understand their world (Broussine 2008; D'Cruz et al. 2021; Guillemin 2004).

This was particularly so in our case because our research was with children of one of India's most marginalised communities – the Bhils – whose language (Bhili) presented a major hurdle in conducting research with them. This was overcome by taking the aid of local activists and using drawings as a research method. We therefore agree with Literat (2013) that drawing succeeds where the language barrier between the researcher and the participant may seem like an insurmountable obstacle that is bound to lead to failure. Beyond this, drawings is a popular way for many children to communicate because they cannot fully articulate their beliefs and emotions using spoken or written words (Guillemin 2004; Tay-Lim and Lim 2013). This was certainly the case with some of our participants who were very young and could not speak for themselves. In fact, despite drawing, some children did not want to speak or spoke in monosyllables. In such a situation, other children or local activists assisted our understanding by prompting them to talk and provide significant accounts of their experiences at work. Of course, all through the process, we tried to stay cognisant of the need to ensure that the children did not feel compelled to share their stories. Given this, if it was not for the drawings it would not be possible to do this research (D'Cruz et al. 2021). The act of drawing allowed participants to feel more willing and able to share hidden parts of their subjectivity that would be difficult to access if we solely relied on interviews (Broussine 2008; Ward and Shortt 2019). Moreover, since child labour had become a sensitive issue, the most appropriate method was drawings. Further, since most of the follow-up interviews were conducted in the presence of elders or those who had an interest in the Bt cottonseed industry, their presence could have constrained the children from talking to us if it were not for drawings. Thus, when combined with interviews it enabled a richer, more complex, deeper and sometimes elicited hidden data such as bullying in the workplace rather than overly simplistic conceptualisations and misrepresentations (Oliveira 2016; Rees 2018).

In this work we use participant-produced images rather than pre-existing researcher-produced visual stimuli. The advantage of using participant-produced images was that it gave the researchers some time to familiarise themselves with children from a different culture who were extremely shy and did not speak the language of the researchers. Further, it also enabled us to implement a participant-centred approach, whereby the researchers actively collaborated with children to produce knowledge and meaning. This also changed the researcher-participant relationships as participant-generated images shifted the balance of power away from researchers to the participants, giving voice to children who are typically silent in other studies on child labour. In fact, we endorse the assumption that children are active, competent and reflexive expert informants about their own lives (Tay-Lim and Lim 2013). It thus made possible for us to place these children at the centre of the research process (Barker and Weller 2003).

We also abandoned the 'draw-and-talk' procedure, though widely advocated (Tay-Lim and Lim 2013), after our experience at the first village. It seemed to be a major hindrance to the process as children got conscious and were hesitant to talk while drawing. The 'draw followed by talk' seemed better because it did not interfere with the activity and children did not feel dominated and supervised. They enjoyed the process when left alone. However, our control over the process did not completely dissipate. We were more direct in our instructions as the study progressed and moved away from the location after the instructions were given, only to return time again to check on the children and answer their clarifications. Nonetheless, the value of listening to children's voices has methodological consequences (Tay-Lim and Lim 2013). Participants played a key role regarding the interpretation of the drawings, which also empowered them in expressing their inner voices. Thus, we strongly agree that drawings are most fruitful when combined with interviews particularly in the case of research with children (Punch 2002).

Further, as was so in our case, researchers should be prepared to expect stories of trauma and personal accounts of pain and suffering and be prepared to handle emotions and feelings in empathetic, sophisticated and considerate ways (Guillemin and Drew 2010; Ward and Shortt 2019). This forms part of the need to be reflective, ethical and building a relationship of empathy and mutual respect with participants (England 1994) particularly in the differential power relationship while researching children (Thomas and O'Kane 1998).

However, using drawings also raises some significant questions of individual skills. Those who perceive themselves to be 'no good' at drawing or writing are likely to feel constrained. Besides this, drawing is a mode of communication with particular 'affordances' or potentialities that are socially defined, this relates to the question raised earlier about how children might perceive the invitation to draw (Buckingham 2009). In our case too, some children looked around and just scribbled, while others took keen interest.

Finally, the concept of empowerment implies that research should also benefit those who are subjected to it and help to solve problems of communities (Pauwels 2015). In our case, this study was done for an NGO working with children in Bt cottonseed farms which had initiated action against child labour (Banday et al. 2018). Though this action can be seen as benevolent, we cannot be sure if this intervention is what the community wanted. Importantly, the research agenda was identified by Prayas Centre for Labour Research and Action and the funder Sudwind Institute, Austria, rather than by research participants themselves. Consequently, whether empowerment is realised through this, or any other community-focused, research project, remains an open question.

Notes

1 Under section 391 of the Indian Penal Code dacoity means five or more persons conjointly commit or attempt to commit robbery.

2 Bhil Corps was a military outfit of Bhil recruits headed by a British officer. With the assistance of the Corps the marauding tendencies of the hill Bhils were suppressed and tranquillity restored. This enabled the more comprehensive efforts to promote sedentary agriculture (Nilsen 2015).

3 Bt cotton, a gene from the soil bacterium *Bacillus thuringiensis*, is genetically engineered into a cotton variety to produce a protein which acts as an insecticide (McKinney 2013).

References

Banday, Muneeb Ul Lateef, Saikat Chakraborty, Premilla D'Cruz, and Ernesto Noronha. 2018. "Abuse Faced by Child labourers: Novel Territory in Workplace Bullying." In *Indian Perspectives on Workplace Bullying*, edited by Premilla D'Cruz, Ernesto Noronha, Avina Mendonca, and Nidhi Mishra, 173–204. Singapore: Springer.

Barker, John, and Susie Weller. 2003. "'Is it Fun?' Developing Children Centred Research Methods." *International Journal of Sociology and Social Policy* 23 (1–2): 33–58.

Bell, Emma, and Jane Davison. 2013. "Visual Management Studies: Empirical and Theoretical Approaches." *International Journal of Management Reviews* 15 (2): 167–184.

Bolton, Angela, Christopher Pole, and Phillip Mizen. 2001. "Picture This: Researching Child Workers." *Sociology* 35 (2): 501–518.

Broussine, Mike. 2008. *Creative Methods in Organizational Research*. London: Sage.

Buckingham, David. 2009. "'Creative' Visual Methods in Media Research: Possibilities, Problems and Proposals." *Media, Culture & Society* 31 (4): 633–652.

D'Cruz, Premilla, Ernesto Noronha, Muneeb Ul Lateef Banday, and Saikat Chakraborty. 2021. "Place Matters: (Dis)embeddedness and Child Labourers' Experiences of Depersonalized Bullying in Indian Bt Cottonseed Global Production Networks." *Journal of Business Ethics*. https://doi.org/10.1007/s10551-020-04676-1

England, Kim V. L. 1994. "Getting Personal: Reflexivity, Positionality, and Feminist Research." *The Professional Geographer* 46 (1): 80–89.

Guillemin, Marilys. 2004. "Understanding Illness: Using Drawings as a Research Method." *Qualitative Health Research* 14 (2): 272–289.

Guillemin, Marilys, and Sarah Drew. 2010. "Questions of Process in Participant-generated Visual Methodologies." *Visual Studies* 25 (2): 175–188.

Harrison, Barbara. 2002. "Seeing Health and Illness Worlds – Using Visual Methodologies in a Sociology of Health and Illness: A Methodological Review." *Sociology of Health & Illness* 24 (6): 856–872.

Lenette, Caroline, and Boddy Jennifer. 2013. "Visual Ethnography and Refugee Women: Nuanced Understandings of Lived Experiences." *Qualitative Research Journal* 13 (1): 72–89.

Literat, Ioana. 2013. "'A Pencil for Your Thoughts': Participatory Drawing as a Visual Research Method with Children and Youth." *International Journal of Qualitative Methods* 12 (1): 84–98.

McKinney, Kacy. 2013. "Troubling Notions of Farmer Choice: Hybrid Bt Cotton Seed Production in Western India." *Journal of Peasant Studies* 40 (2): 351–378.

McKinney, Kacy. 2014. "'Hybrid Cottonseed Production Is Children's Work': Making Sense of Migration and Wage Labor in Western India." *ACME: An International E-Journal for Critical Geographies* 13 (2): 404–423.

Mosse, David, Sanjeev Gupta, and V. Shah. 2005. "On the Margins in the City: Adivasi Seasonal Labour Migration in Western India." *Economic and Political Weekly* 40 (28): 3025–3038.

Nilsen, Alf Gunvald. 2015. "Subalterns and the State in the Longue Durée: Notes from 'The Rebellious Century' in the Bhil Heartland." *Journal of Contemporary Asia* 45 (4): 574–595. https://doi.org/10.1080/00472336.2015.1034159.

Oliveira, Elsa. 2016. "Empowering, Invasive or a Little Bit of Both? A Reflection on the Use of Visual and Narrative Methods in Research with Migrant Sex Workers in South Africa." *Visual Studies* 31 (3): 260–278.

Pauwels, Luc. 2015. "'Participatory' Visual Research Revisited: A Critical-Constructive Assessment of Epistemological, Methodological and Social Activist Tenets." *Ethnography* 16 (1): 95–117.

Phillips, Maxwell P. 2012. "Dialect Continuum in the Bhil Tribal Belt: Grammatical Aspects." PhD diss., University of London.

Punch, Samantha. 2002. "Research with Children: The Same or Different from Research with Adults?" *Childhood* 9 (3): 321–341.

Rees, Charlotte. 2018. "Drawing on Drawings: Moving beyond Text in Health Professions Education Research." *Perspectives on Medical Education* 7 (3): 166–173.

Roche, David. 2000. "The 'Ḍhāk', Devi Amba's Hourglass Drum in Tribal Southern Rajasthan, India." *Asian Music* 32 (1): 59–99.

Ross, Karen. 2017. "Making Empowering Choices: How Methodology Matters for Empowering Research Participants." *Forum: Qualitative Social Research* 18 (3): 128–144.

Sinha, Bobby Luthra. 2017. "Through the Looking Glass of Bhils – How Markets Win What the State Has Lost in the Desert Jungles of Western Rajasthan." *Journal de Ciencias Sociales* 9.

Sjoi̇̈blom, Disa Kyllikki. 1999. "Land Matters: Social Relations and Livelihoods in a Bhil Community in Rajasthan, India." PhD diss., University of East Anglia.

Skaria, Ajay. 1997. "Shades of Wildness Tribe, Caste, and Gender in Western India." *The Journal of Asian Studies* 56 (3): 726–745.

Skaria, Ajay. 1998. "Being Jangli: The Politics of Wildness." *Studies in History* 14 (2): 193–215. https://doi.org/10.1177/025764309801400203.

Tandon, Indrakshi. 2019. "'We Get Nothing': An Ethnography of Participatory Development and Gender Mainstreaming in a Water Project for the Bhil of Central India." PhD diss., State University of New York at Albany.

Tay-Lim, Joanna, and Sirene Lim. 2013. "Privileging Younger Children's Voices in Research: Use of Drawings and a Co-Construction Process." *International Journal of Qualitative Methods* 12 (1): 65–83.

Thomas, Nigel, and Claire O'Kane. 1998. "The Ethics of Participatory Research with Children." *Children & Society* 12 (5): 336–348.

Ward, Jenna, and Harriet Shortt. 2019. Drawing. In *The SAGE Handbook of Qualitative Business and Management Research Methods: Methods and Challenges*, edited by Catherine Cassell, Ann Cunliffe, and Gina Grandy, 262–281. London: Sage.

White, Allen, Naomi Bushin, Fina Carpena-Méndez, and Caitríona Ní Laoire. 2010. "Using Visual Methodologies to Explore Contemporary Irish Childhoods." *Qualitative Research* 10 (2): 143–158.

6

CREATIVE MEMORY

Memory, methodology and the post-colonial imagination

Jasmine Hornabrook, Clelia Clini and Emily Keightley

Introduction

In this chapter we introduce the creative methods employed in the Migrant Memory and the Post-colonial Imagination project (MMPI), a five-year research project focusing on the remembering of the 1947 Partition of British India and post-colonial migration by South Asian communities in the UK. We do so in order to show how creative participatory methods provide solutions to methodological challenges in researching memory, not least those related to doing research on memories of painful or marginalised pasts. Research of this kind requires the creation of opportunities for participants to approach feelings of belonging, discrimination, isolation and marginalisation in oblique ways and provide safe expressive spaces for their articulation. We argue that creative methods contribute to the collaborative production of knowledge about community and belonging in South Asian diasporic communities. We begin by introducing the project and our fieldwork sites before addressing some of the methodological challenges of the project. We then discuss our methodological approach that combines ethnographic tools with creative community activities, before presenting our creative methods in practice. While the use of creative methods poses challenges, we argue that the use of such methodologies, in a reflexive and responsive way, has the potential to develop more egalitarian research processes.

Creative methodologies use cultural participatory activities that encourage participants to creatively engage with research themes through cooking, textiles, photography, film and music. These methods are creative in the sense that the research activities are culturally generative, producing both intangible and tangible cultural forms through the research process, rather than being simply extractive. These methods have the potential to evoke and elicit unique data by providing alternative ways to articulate abstract concepts, implicit mnemonic meaning and

DOI: 10.4324/9780429352492-6

individual and collective identities in comparison to standard ethnographic tools (Chakraborty 2009; Beebeejaun et al. 2013; Harper 2002; Tolia-Kelly 2004; van der Vaart et al. 2018). Creative methodologies also provide space for the articulation of personal and cultural memories and in doing so involve the active production of community and personal relationships and identities. More broadly, such creative methods can also promote shifts in the power imbalances within the research process through the promotion of collaboration, participation and inclusion of multiple claims to knowledge. We argue that reflexivity and responsiveness in creative methodologies within a longitudinal, multi-perspective study is imperative to achieving two of our central aims. Firstly, to create collaborative and more egalitarian understandings of post-colonial community and belonging in relation to memories of Partition and migration. Secondly, to develop more nuanced understandings of the diversity of roles that different everyday media and creative practices play in the communication of difficult pasts.

Partition remains a sensitive and painful historical event, remembered differently by different groups and at different scales, from the individual to the collective. Our research design and methodology, therefore, needed to take into account the multiple perspectives and multiple ways of knowing that we encounter in diverse societies. Memory plays a significant role, meshing with collective consciousness and self-identification "to shape a sense of belonging and affiliative membership within a real or imagined community" (Kearney 2013, 132). The concept of 'community' is highly complex and multi-layered, often with porous ethnic boundaries that can speak against ideas of "reified culture" (Baumann 1996; Williamson and DeSouza 2007). Similarly, a sense of 'belonging' can overlap numerous social locations, individual identifications and emotional attachments and ethical and political value systems (Yuval-Davis 2006). Finally, 'home' is plural, affective, subjective and not necessarily utopian (Ahmed 1999), and, in diasporic contexts, it can refer to a 'homing desire' rather than the homeland itself (Brah 1996). In the contested histories of communal politics, tension and violence that Partition encapsulates and the multiple migration trajectories and experiences of settlement in the UK, collaborative and multiple understandings of community and belonging become all the more important. Creative, participatory methodologies provide the space for the type of reciprocity and feedback (Tuhiwai Smith 2012; see also Beebeejaun et al. 2013; Feld 2012; Williamson and DeSouza 2007; Ross 2017) that lead us towards more nuanced understandings of 'community' and 'belonging' in South Asian diasporic communities in our ethnographic sites of Loughborough and London.

Given the post-colonial focus of the project, it is crucial to analyse the power imbalances involved in knowledge production itself (Foucault 1981), to consider the connection of research to European imperialism and colonialism (Clifford 1986; Fabian 2012; Tuhiwai Smith 2012), and to be "accountable to the past" (Gunaratnam 2003, 7). Linda Tuhiwai Smith (2012) argues that, for the colonised, the collective memory of imperialism is perpetuated through the ways knowledge has been collected, classified in the West and is then represented back to those who have been colonised, in a process of interchange between scholarly and imaginative construction of ideas about the Orient (see Said 2003; Clifford 1986). Both the formal scholarly pursuits of knowledge and the informal, imaginative and anecdotal constructions of the Other are intertwined with each other and with the activity of research; therefore, researchers need to understand and analyse the complex ways in which the pursuit of knowledge is deeply embedded in the multiple layers of imperial and colonial practices (Tuhiwai Smith 2012, 2). Shifting research frameworks and methodologies to become more open, creative and reciprocal with particular, often marginalised, communities (14) is therefore central to the process of problematising these taken-for-granted perspectives. The shift towards co-production in research moves towards more equal partnerships with communities and researchers; working together in collaboration, to understand issues and create knowledge; accepting different claims to knowledge; and more reflexivity in assessing the positionality of the research and the researcher (Beebeejaun et al. 2013, 41; Enria 2016; Tolia-Kelly 2007). Such a shift can be facilitated through the co-design of cultural research activities with communities, through using creative tasks as alternative ways through which to discuss community and belonging and to disseminate knowledge through community as well as academic outputs and seeing these two forms of knowledge dissemination as mutually informing one another. While we aspire to create space for more egalitarian research processes, we acknowledge the limitations on the co-production of knowledge and the almost insurmountable challenge of completely removing power imbalances in research (see Enria 2016; Ross 2017). As researchers, we have research agendas and, with such agendas, we realise the limitations of eradicating power imbalances. This chapter therefore presents a self-reflexive account of our methodological approach in the MMPI project and considers the role of the research and researchers in relation to communities, in working *with* communities, rather than *on* communities. In doing so, we aim to collaborate and collate multiple perspectives on community and belonging in British South Asian diasporic communities in our research.

Migrant memory and the post-colonial imagination project

MMPI is a five-year project based at Loughborough University and funded by the Leverhulme Trust. The project investigates: what memories of Partition circulate within South Asian communities living in Loughborough and London; how such memories are communicated over time and space; how social practices and processes of remembering Partition inform intercommunal identities and the (re) construction and the idea of community itself; and what role these memories play in the articulation of a sense of identity and belonging, both in relation to British Asian identities and also to a sense of Britishness in contemporary social life. We are particularly concerned with understanding how the past shapes present British South Asian inter-community relations in a post-colonial context in the specific localities of Loughborough and Tower Hamlets. As Sean McLoughlin (2013, 16) suggests, "simple recollections of the stuff of everyday life can unlock the placed, embodied and affective maps of memory and identity which are at the heart of diasporic consciousness". By excavating inherited, cultural memories of Partition as part of a longer process of decolonisation, along with autobiographical memories of migration, we gain an understanding of how these pasts are involved in the social construction of identities in diasporic settings.

Memories of historical events such as Partition can be fragmented and partial and memory studies as an analytical approach attends to and interprets the meaning of these mnemonic pieces. Ananya Jahanara Kabir (2009) critiques Partition studies for an over-reliance on narrative as the favoured mode for creative practice and academic scrutiny, with the focus on historiography, anthropological investigation and literary and cultural studies overwhelmingly on primary materials that have strong narrative components, such as the short story and novel, oral history and popular cinema (p. 489). Given the unfinished character of narratives living on in different versions of the social memory of different social groups (Das 1998, 118; Feuchtwang 2000) and the potential for excessive narrativisation of Partition, Kabir suggests that analysis should focus on embedded non-narrative moments, such as theatre, music, lyric poetry, photography, painting, sculpture, and public monuments (2009, 489–490). Deploying creative participatory methodologies as part of a memory studies approach rather than in an oral history framework is a way of accounting for non-narrative elements of the past in the present. Using the mnemonic fragments elicited from creative processes within our community activities is central to gaining understandings of the social impact of Partition memory. Following Passerini, Radstone (2000, 10) suggests that, in relation to oral history, some historians have

> been struggling to hold in tension an understanding of oral history testimony that acknowledges its *relation* to "happenings", to "the constituted", to historical experience while developing an understanding of memory "as an active production of meanings and interpretations, strategic in character and capable of influencing the present" (Passerini 1983, 195).

Memory is therefore understood as "a text to be deciphered, not a lost reality to be discovered" (King 1997, 62), with representations of memory understood in relation to both cultural narratives and unconscious processes (Radstone 2000, 10). In relation to Partition, this is at least in part a methodological issue that can be addressed through the use of methods which engage and employ everyday creative processes of remembering Partition, Empire and migration in contemporary Britain. Such memories of Partition and its aftermath are further complicated for South Asian diasporic communities in the UK. In South Asia, the language of nationality has been tied to and created with the ideologies and power structures of colonialism, through the assignment of the British administrative category of India and the subjectivity as 'Indian' manifested in anticolonial nationalism (Shukla 2001, 560). Indian, Pakistani and Bangladeshi nationalities are entwined within histories of colonialism, communalism and conflict. Therefore, multiple perspectives and ways of expressing national identity become all the more important and creative community activities allow for alternative articulations of nationality in a variety of ways. For instance, during two different activities with different groups, national identity was expressed through material culture. One participant expressed Indianness through wearing a *sari*:

> (What does wearing a sari say about you?) I am an Indian lady and I've come from India and I would like to stick to that. I would like to stick to my identity, who I am and who we were, what we were and things like that, yes. Very, very proud.

Another participant expressed Bangladeshi identity through food:

> Hilsa fish is our national fish . . . baate masse Bengali (rice and fish makes you Bengali).

To gain multiple perspectives from individuals and groups from diverse South Asian backgrounds, the MMPI project is a longitudinal study based on immersive fieldwork based in two sites in Loughborough in the East Midlands and in Tower Hamlets, London. Experiences and memories of Empire, Partition and migration vastly differ among participants and between the two fieldwork sites that offer a contrast in terms of size, demographic, region and urban environments; the market town of Loughborough is in the borough of Charnwood has a population 166,100 people (NOMIS 2011) whereas the borough of Tower Hamlets in East London, with a population of 317,200 (Tower Hamlets Council 2018). Despite its size, Loughborough is home to a diverse population, with a higher-than-average South Asian demographic of 12,675 people of Indian, Pakistani and Bangladeshi background (NOMIS 2011). Tower Hamlets, on the other hand, is well known for its large Bangladeshi population, which makes up the 32% of the population, while a 3% is Indian and 1% is Pakistani (Tower Hamlets Council 2013). Many of the community groups with whom we work have lived in the UK for decades, and have acquired British citizenship; others were born in the UK. Many identify as 'British Asian' and/or 'British Indian', 'British Bangladeshi', 'British Pakistani' and/or 'British Muslim'. As Sandhya Shukla (2001, 551) asserts, multiple formations of nationality take place in diasporic culture, putting into question the geographical and conceptual boundaries of community, which are complicated further by their localised manifestations in diverse sites such Loughborough and Tower Hamlets, in which lived, everyday experience overlaps not one but many diasporas (Shukla 2001, 563; see Baumann 1996). It is therefore vital to gain nuanced and diverse understandings of community in our work.

Our diasporic fieldwork sites are further complicated by the diversity and complexity of memories and processes of remembering surrounding Partition across different social, cultural and religious categories. For instance, we work with numerous community groups and organisations from women's meeting groups for over 65s, well-being groups counteracting isolation amongst Bangladeshi Muslim women, English-learning groups, cultural organisations, community members and devotees of Hindu temples and Gurudwaras and networking organisations for professionals. The diversity of the participants in terms of generation, religion, gender, class and caste reflects the diversity of memories and narratives of Partition that have been circulated, mediated and inherited across time and space. Such intersections can also profoundly shape the 'diaspora space' (Brah 1996) and impact how 'British' and 'Asian' identities are negotiated and how belonging is experienced (Bakrania 2013). The maintenance of significant transnational connections – through familial, social and cultural networks and imaginaries – in a globalised world also informs conceptions of community as social formations "come into being through imaginative and political renderings of themselves elsewhere" (Shukla 2001, 564). Therefore, a variety of different memories, knowledges and accounts of community, belonging and identity emerge from different social groups within the fieldwork sites. The diverse accounts, memories and understandings of Partition, migration, community and belonging in South Asian diasporic communities lie at the centre of our research and, we argue, creative methodologies create the necessary space for multiple perspectives.

Methodological challenges and approach

In conducting research on the role of Partition memory in the experience of identity and belonging in contemporary diasporic communities, the MMPI project faces a number of methodological challenges. While the vast majority of our participants did not directly experience Partition, intergenerational, inherited memories and widely circulating cultural memories are key to understanding the complex, multi-layered ways in which remembering is socially experienced, practiced and performed across time, space and different social categories (Pickering and Keightley 2013, 105). Addressing the specificity of second-hand memories in terms of their residual connections to an unexperienced past and their implicit articulation and explicit mobilisation in the present can be difficult to tease out using conventional ethnographic methods.

Investigating long-held memories of Partition across a wide range of individuals and groups is difficult in terms of the pain evoked by questioning and the ethical implications of re-remembering communal tension for the purposes of research. In an interview, for instance, a first-generation British Indian woman[1] responded to the question "When do you talk about Partition?" by answering: "It's too painful. . . . We don't talk about Partition. We just don't talk about it. . . . It's gone, it's gone, because it was '47, 1947, so we don't talk about it. . . . It's a long time now, but you have to move on" (personal communication, June 2019). Both managing and acknowledging the importance of pain in the research process can be a delicate task, and a failure to do so can disrupt encounters with research participants and even relationships with whole groups. Ethnographic methods such as one-to-one interviews provide a narrow range of opportunities for dealing with pain and so alternative modes of exploring painful memories in ways that can accommodate pain without being overwhelming were needed.

The spatial dimensions of Partition memory mean that for many of the participating individuals and community groups in the UK, remembering practices and processes traverse across continents both then and now. Many participants are geographically removed from the physical site of Partition and live outside the nations that were created as a result of the division; however, strong connections are maintained through family and national imaginaries. Accounting for the transnational nature of remembering processes can be difficult using locally embedded ethnographic methods.

The project took a number of approaches to tackle these methodological challenges. The research framework itself was designed with community collaboration as a central feature. In both fieldwork sites, we work in conjunction with community partner organisations to co-design, conduct and disseminate the research. In Loughborough we work with Charnwood Arts – an independent community arts and media organisation working with community groups – and Equality Action – an organisation that works with communities on matters of equality and immigration and runs projects aimed at marginalised residents, while in London we work together with the arts and culture office at Tower Hamlets Council. These partnerships are vital in both identifying and recruiting participants and community groups who are interested to work with us and in co-designing creative activities that are culturally safe, appropriate and relevant to participants' experiences and memories.

Our methodological approach centres on a combination of ethnography and creative community activities. Modes of ethnographic inquiry include interviews, focus groups, group discussions, participant observation and field notes.[2] These ethnographic methods, particularly interviews and focus groups, provide an opportunity to focus directly on our research questions which inevitably impose the intellectual agenda and analytical concerns of the researchers onto these interactions. However, creative participatory methods, such as food and memory sessions, fashion and textile projects and photography workshops, are then used to open up spaces for multiple claims to knowledge to be made and to challenge and question the meanings of community, memory and post-colonial experience that the researchers have used. These claims and challenges then inform interactions in interviews and focus groups. This dual approach combines the naturalistic features of ethnography by embedding the research teams in the communities they are researching with the active creation of spaces in which remembering can be explicitly articulated and explored. This combination is participatory rather than extractive in the sense that we work in, and with, community groups to design the creative spaces for remembering. It is also responsive in the sense that one-to-one interviews, group discussions and creative activities produce ideas and suggestions for new activities, and is reflexive insofar as the role of the researcher as a designer, facilitator and participant in creative activities continually requires negotiation. Through this multi-method approach, we aim for triangulation of data

(Chakraborty 2009, 422) that accommodates multiple perspectives on community and belonging.

Creative participatory activities backed up by ethnographic inquiry enable us to explore the movement of memory across the social scales – from the individual to the collective, the private to the public, the cognitive to the social (Keightley and Pickering 2013, 9). With the potential of creative methods in both excavating memories and creating more dialogical spaces in the research process, we now provide a self-reflexive overview of the creative methods and community activities we employ.

Creative methods in Loughborough and Tower Hamlets: articulating memory, evoking 'community'

The starting point of the project's community activities was the screening of Bollywood and British Asian films in a focus group setting, followed by a group discussion. Participants for the first focus groups, in both Loughborough and London, were initially recruited via our partner organisations: Equality Action and Tower Hamlets Council. We chose film screenings and discussions for two primary reasons. First, we were cultural outsiders and outsiders to the fieldwork sites. Through standard ethnographic methods, significant time had been given to immerse ourselves in the field sites through attending group meetings and cultural and religious events to gain familiarity with group members, to build rapport and trust and to observe the context of, and participate in, cultural practices, such as the Loughborough and London Mela, The Season of Bangladrama and other public events of significance for the communities (poetry, music, theatre, literary or religious events). However, we used the screenings to introduce the project and to build relationships with groups. Second, we wanted to start with an activity that would tap into existing cultural practices to engage with, and appeal to, participants. Considering how, as Bissell et al. (2000, 170) posit, "film can provide a provocative medium for allowing its subject an opportunity to interpret, question and theorise" but, at the same time, they can be "not only evocative but also entertaining" (174), film screenings proved to be a good starting point. Given their popularity amongst South Asian diasporic communities (Dudrah 2006; Puwar 2007; Sardar 1998), screenings of Bollywood films in particular, and other films focused on colonial South Asian history, were organised. These film-based activities enabled a first stage of embeddedness into community groups, to raise interest from other individuals and groups, and to create an initial dialogic space for thoughts and memories in response to the films. Films such as *Viceroy's House, Veer Zaara, Lagaan, Raazi* and *Midnight's Children*, which focus on colonialism, Partition and its aftermath, provided a platform for reflections on memories and postmemories of Partition, as participants actively made connections between family memories and what they saw on screen. For example, commenting on *Viceroy's House*, a participant in London observed:

> I grew up with my grandma telling me all these stories of how she used to hide and stuff when there were Hindus coming, and they found all these

different ways of just staying from conflict altogether and seeing that just made everything that she told me come to life even more as well.

Such screenings were also very productive in gaining a preliminary understanding of what was (un)known about Partition. Many participants criticised the fact that Partition is not part of the British national curriculum, and, even though they appreciated that the film portrayed fictionalised accounts of the events, they valued the fact that the films give visibility to a history that has been told only within certain families or has not been told at all. Discussions quickly moved away from films to focus on family memories and personal reflections on the themes which emerged through film narratives. In some cases, considerations on one's family history also extended to political reflections on colonialism and its legacy in post-colonial Britain:

> These people on the pedestal almost and literally just ignore anything that they've done that's bad and . . . that's the whole other narrative that we also hear a lot, like we hear the whole thing about how much like India, Pakistan, Bangladesh have suffered, but equally we hear like whenever those discussions come up, like we hear those kind of voices going, "Oh, but yeah, the British gave us railway," or, "The British gave us this and that".
>
> I mean there's still that colonial mindset almost where we're like, "No, still we're going to have to respect them". It's kind of like you know when the master comes in and literally everyone's like bowing their heads and all of that, we still have that mind set almost, or maybe not us as in our generation, but our parents' generation still – in the same breath they're complaining but at the same time they're also kind of having that mentality where they're like – whatever they've done, we're still inferior and we should respect.

As an initial series of activities, with the aim of introducing ourselves in the social landscape of community activities in both sites, these screenings provided a good way to begin conversations surrounding our research theme. Moreover, these sessions were also valuable in identifying and recruiting willing participants for interview and to set the stage for the organisation of future activities with a range of different community groups.

Having gained some understanding of the groups and of the social dynamics and cultural landscape of the two locations through our immersion in the local sites and through the film screenings, we then started to approach specific groups with whom we co-designed creative activities that integrate and build on the lived experiences, personal memories and cultural and creative practices of our participants. By referencing and reflecting the histories and creative practices of the groups with whom we work, we work towards attaining a trustworthy rapport through familiar actions, evoking sensory memories and accumulating tangible accounts of memory (Tolia-Kelly 2004, 4). For instance, many of the Bangladeshi participants work, or have worked, in the catering industry in 'Indian' restaurants in the UK, while many participants in Loughborough worked in the textile industry as factory workers for much of their working lives. In response to this, food and memory cooking sessions were co-designed with Bangladeshi groups in Loughborough and London and a fashion and textiles project with a Gujarati women's group in Loughborough.

Food and memory

In the food and memory cooking sessions in Loughborough we worked with a predominantly Bangladeshi men's group and Bangladeshi women's group. Both were established community groups led by Equality Action, which served to raise awareness of mental health issues in marginalised communities and to tackle social isolation. The women's group predominantly consisted of first-generation women who migrated as a result of marriage to partners already settled in Loughborough. A smaller number of women migrated from Bangladesh via Italy, were born in Bangladesh and brought up in the UK or were born in the UK. The majority of the women were practicing Muslims. Relationships were built with the groups through regular visits to their meetings, getting to know group members and through film screenings. We had numerous discussions about the possibilities of the cooking sessions with the group, contemplating what we could do and how memories can be recalled through food. Co-designed with our research assistants, who also work for our partner organisation and facilitate the groups, and with the groups themselves, three sessions were organised in each series. We worked on specific themes for each session to aid with the selection of the dishes and to direct the types of memories to be explored during the group discussion. Themes included forgotten dishes, food from childhood and food from 'home', and we encouraged open interpretation of these themes. A similar series of food and memory sessions was also organised in Tower Hamlets in collaboration with Mulberry School for Girls. As part of its engagement with the local community, the school organises a series of group activities that target mothers, with the aim of promoting a sense of community and the creation of a network for women who might not have the time or the opportunity to develop other personal relationships. A structure was agreed with the Parents' Liaison Officer, who also recruited the sessions' participants, and followed a similar set of themes to the Loughborough sessions.

Prior to each of the cooking sessions in Loughborough, the groups decided on Sylheti, Bangladeshi and some Italian dishes, reflecting participants' everyday cooking practices, places of origin and trajectories of migration. The participants took turns to decide what dishes to cook and to lead the cooking sessions, and, amid the sounds and smells of the kitchen, we facilitated several rounds of group discussion and short one-to-one discussions based on the dish and theme of the session. Once the dish was ready, the group would eat together and respond to questions based on the dishes, using the dish to evoke sensory memories as well as triggering personal and cultural memories and reflections of community and belonging. The participant-centred activities became a space for participants to express themselves and to enter into dialogue with others in different ways, from the decision-making surrounding the dishes for each session to correlate with the themes, the creative act of cooking the dishes and the discussion focusing on food and memory.

The sessions proved productive in evoking personal and cultural memories as well as providing an alternative space and means to articulate identity and belonging. Throughout the sessions, participants maintained that food strongly connects them and their families to Bangladesh as change over time shifts identities towards a dynamic and postmodern sense of British Islam (see Hoque 2015, 3). While all the sessions were significant in asserting collective Bangladeshi identity and community, the group explicitly articulated Bangladeshi-ness through the connections made between 'home' and the decision to make hilsa fish during the 'food from "home"' session (see Figure 6.1). A collective Bangladeshi-ness was articulated through the choice in making this dish, with a participant immediately starting the conversation with:

> Hilsa fish is our national fish, I think most Bengalis like it . . . I think most Bengalis like their traditional food (I: Is this an important dish all over Bangladesh?) . . . The whole country like hilsa.

The significance of hilsa and other fish was a central feature in discussions of Bangladeshi-ness through food. The saying *"baate masse Bangali"* ("rice and fish makes

FIGURE 6.1 Food and memory workshop, Loughborough. Egg bhuna and hilsa fish. December 2019

you Bengali") was articulated numerous times throughout the activity series in response to questions about the significance of food and food practices in Bangladeshi communities. Religious identity was also asserted through food and the practice of cooking for others, with a participant asserting, "[a]ctually the Muslim people like to cook and eat and feed other people . . . we are famous for that".

An explicit sense of community was evoked through the group creative activity itself, as well as through the active process of remembering. Memories of cooking for and eating with large groups of family, friends and religious leaders in Bangladesh were often reflected on during the sessions. During two of the sessions, a participant expressed how the taste of the dishes improved as she was eating in the 'community' setting of the session. She said:

> I love that kind of food because lots of people [are eating] it. I'm thinking it's more tasty. I don't know why. Is it my memory or my mind, I don't know. I find it's more tasty.
>
> Sometimes I made it [the dish at home], it's okay, but I want like [a] community.

While the participants cook at home every day, the experience of attending the sessions as a group evoked memories of 'community' in Bangladesh. In particular, the experience of coming together to participate in social cooking and eating in the sessions reminded the participants of their personal memories of the Bengali cultural practice of *tufabat*, in which young people get together to cook and eat a dish outdoors in Bangladesh. Memories of *tufabat* became a frequent topic in our discussions, which led to reflections on the change of such practices as a result of migration and resettlement. The groups and their children do not practice *tufabat* in Loughborough and the dishes and the practices of social cooking and eating during the session provided a focus point to reflect on cultural change across the generations. This particular memory and the focus on change in Bangladeshi food practices provided space for the participants to articulate their concerns over the continuation of Bangladeshi culture in Loughborough. For instance, discussions around passing these recipes down to their children elicited a mixed response but led to assertions that their children are assimilating to "become more and more British" through their food practices. As food is a significant marker of cultural continuity, as well as difference, hybridity and/or assimilation (Mankekar 2015, 84), discussions around dishes allowed us to cover such discussions of cultural continuity and assimilation and to open up reflections around differences and similarities with neighbouring communities. For instance, reflecting on the differences of Indian and British food in comparison to Bangladeshi food, the participants discussed:

> Yeah, like fruit and vegetables are the same [in India] . . . (Chapatis, rotis, same.) But the way of cooking is a little bit different, isn't it? Little, spice and – more spices or different ways. And curries are different. Curries, yeah,

they use kari patta [curry leaves] all the time. We use, after cooking we put dhania patta – coriander leaves. So they are a little bit different but similar.

I think the British learn from us, lots of curries and things they want now, and spices. We're eating your food as well . . . like roasts and things, we make them our way with the spices, but British is not spicy, isn't it? . . . So they learn from us, the spice, they bring the spices, rice, everything here now, Indian food, restaurants are popular here, isn't it?

Reflections on food practices is a tangible and relatable way of articulating concepts such as cultural difference and hybridity and reveal how participants position themselves in relation to other nationalities (even if they are themselves British citizens). The focus on food and cooking also countered concerns over memory that many of the participants voiced. In London, for instance, some of the participants were initially reluctant to participate over concerns that they would have not been able to remember very much. In the pilot session, during which we illustrated the project to participants, a trip to the food market was followed by a very lively discussion on fruit in Bangladesh (see Figure 6.2) and how they connected the taste and smell of fruits with episodes of their childhood, or to their visits to Bangladesh:

So this is one of my favourite fruits. It reminds me of my childhood. My mum used to always peel it for us.

I remember a memory when I went to Bangladesh after I got married and when they came to pick me up from the airport, they bought these lychees and it was on the stem, with the leaves and everything and it was so tasty. I've never tasted anything like it, because we had it in this country and the flavour isn't as pungent as these over there.

FIGURE 6.2 Food and memory workshop Tower Hamlets, London. Preparing pomegranates. December 2019

The sensory experiences of the session allowed for the invocation of memories, feelings, histories, places and moments in time (Choo 2004, 209), in a dialogic and collective way. After the recollection of autobiographical memories through food in the first session, the participants literally took the lead in the organisation of the following three workshops. Similarly, in Loughborough, cultural memories of historical events such as the Bangladeshi Liberation War and the 1974 post-independence famine were evoked and talked about through discussions around food practices. For instance, discussions around the use of ingredients in Bangladesh and in British Bangladeshi households led to reflections of the resourcefulness of ingredients due to the famine of post-independence Bangladesh in 1974:

> The seed is so tasty. No Bengalis ever throw away the seeds, but some people buy the jackfruit, the whole thing, just for the seeds (laughs).
>   Jackfruit leaves, she cut them and made pakoras. I never knew that, I knew the goats and cows eat them (laughs), we can eat them as well. . . . Actually, to tell you the truth, when it's things like famine and things like that, people eat anything. They'll eat anything in the bin even, it's so sad, but it's true. It happened in 83 or something, after our liberation, there was a big length of time. So there's a few films about that, it's really sad. Like they will go to the pond and look for food, little fish or anything. They'll dig out the soil and find some root vegetables and things like that. Leaves, they will eat them.

The food and memory sessions opened up a repertoire of memories of Bangladesh and its history, of migration, family life and stories, thus confirming that "food is central to defining the manner by which people's emotional, psychological, social, economic, political, historical, and cultural realities are embodied as a lived and living history" (Abarca and Colby 2016, 4; Carolan 2011). With a broad range of knowledge and experience of food and cooking among the participants, such an activity has the potential to provide a more democratic and inclusive means of participation and articulation of community and belonging. While we moved the explicit focus away from our research themes to memories through food and cooking, the groups discussed national and religious identity, autobiographical and cultural memory of Bangladesh and changes in food practices and community over time and space. Traumatic topics and emotional and symbolic aspects of people's experiences might not be accessed through mainstream methods based on people's verbal or written competence (van der Vaart et al. 2018, 2) but through such activities, dialogue and the act of making facilitates specific cultural practices, knowledge and the senses that provide alternative platforms to express concepts of community, belonging and identity.

Fashion and textiles

A series of activities around fashion, textiles and memory was co-designed with a predominantly Gujarati women's community group in Loughborough. The project was co-designed with the group leader and discussed within the group itself, which consisted of mostly first-generation Indian and East African Indian women. While group members had been concerned over their lack of Partition memory and lack of confidence in speaking about issues of migration, the focus on textiles was an engaging and relevant way to talk through memories and belonging. Significantly, the focus reflected the group's knowledge and lived experiences of sewing at 'home' in Loughborough, East Africa and India and their livelihoods in Loughborough's textiles factories. Creative activities rely on different ways of knowing and therefore potential participants who would not feel comfortable in speaking explicitly about historical political events, such as Partition and the 1971 Liberation War and issues of migration, feel comfortable in sharing autobiographical memories of migration in the context of a familiar and creative act. With the voices of South Asian women historically absent in accounts of Partition in the subcontinent,[3] these creative activities provided a means to access South Asian women's voices in particular which may not have happened if we had maintained use of conventional ethnographic tools. Such creative activities afforded us the potential to reach a range of voices and perspectives across society, not only those of key individuals, community leaders and gatekeepers that are readily accessed through qualitative interviews.

The fashion and textile activities ranged from 'show and tell' sessions with items taken from 'personal textile archives' (Lerpiniere 2013); visits to Gujarati textile collections at New Walk Museum in Leicester (see Barnes and Kraamer 2015); a stitching workshop; and the curation of a 'material memories' fashion show during the annual Loughborough Mela, which led to the co-curation of a museum exhibition with Leicestershire County Council. While several participants from this group have declined to take part in a formal interview, they have contributed to group discussions during these sessions. In such instances, the activities have proved vital for participants to contribute to the research process and for us to access and integrate these multiple perspectives.

During the sessions, the group articulated that fashion and textiles are a significant means to maintain tangible, material connections with India, to articulate 'Indianness' and to engage younger members of their families in South Asian heritage. As vehicles of memory, textiles were particularly significant in the evocation of autobiographical memory and experiences of isolation. For instance, a participant used her mother's intricately self-decorated *saris* as a means to reflect on the 'culture shock' she remembers upon her arrival in the UK and feelings of isolation, having moved to an area without a Gujarati community:

> [P]eople don't have time these days to do embroidery and things like that. Because in Africa and India, there was plenty of time . . . I found it very difficult when I came here in the beginning. Because it was a culture shock for me, a culture shock for me. I'm settled, well settled now. But in the beginning it was very difficult, it was all the time dark, cold and we came in winter, you know, and I didn't find any of my people. . . . And I said to my husband "I don't want to live here". And he said why, and I said, "we don't

have anyone here, it's just you and me. I don't like here, I don't want to live here, to be honest, I'm scared of living here". Then my husband called up a friend in Loughborough, he phoned him on the Wednesday and on the Saturday he came to collect us.

Material objects and the act of stitching became the point of focus and vehicles for memory that allowed for the articulation of painful pasts in more oblique ways (see Figure 6.3). While a participant was initially worried about what she could say when I asked to have a short discussion with her during a session, she quickly went on to reflect on her memories of her multiple migrations between Kenya, India and England following the question "Have you used a sewing machine before?":

I remember a little bit, not much. It's funny, your past, you remember so much when you dig it out. . . . We didn't know nothing about India. When we arrived in a trolley, big truck, we had lots of stuff, we had a bed and a sewing machine and everything. It's a big ship. About fifteen days travelling on the sea. Me and my brothers were young. So when we went in the village with the truck, all the luggage and all the stuff, all the village come out and just staring (laughs), who is it? Somebody come from outside, another country, Kenya, outside, it was something different for them in the village. . . . They would come together around the truck and just looking at us. And we knew nothing about India, nothing at all. . . . It was a strange place for us. Five years we lived there. Slowly, slowly we settled down. Then time to come to England, my father bought house here in England, and he had money, then he just called us.

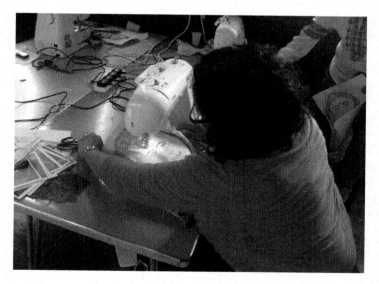

FIGURE 6.3 Stitching Stories workshop, Loughborough. Sewing a self-portrait. July 2019

The focus on different ways of knowing through autobiographical memory and Indian material culture also opens up a space where the roles of researchers and participants go into flux, where participants share significant cultural knowledge that can be reintegrated into the ways we pose our research questions, particularly in interviews and subsequent group sessions. During a visit to New Walk Museum in Leicester, for instance, a participant reflected on a permanent exhibition, which incorporated Gujarati textiles, Hindu deities and Indian toys:

> (I: What was your favourite thing to see in the exhibition?) Oh, well we all discussed that because they've got some gods and goddesses there and everybody said they've got our gods and goddesses here which is really good. Okay, they've gone from A to Z, everything is there, I'll show you, which we were very proud that it will be there all the time. See the top one is Lord Rama, his younger brother Lakshman and his wife Sita-ji and then Hanuman. That's Lakshmi, Narayan, that's Shiva that's Ganesh and there's a Ganesh there. That's baby Krishna and that one we worship that every Monday, you know the Shiva linga, we always . . . do the ceremony by pouring milk, and water and then sandalwood powder and other powders and flowers and all these things. So we all felt that although this museum is in England, they've got our stuff as well.

Here the participant responded to the curation of a museum exhibition that incorporated Hindu deities and, in doing so, articulated a sense of (not)belonging through the inclusion of "our gods and goddess" in the exhibition, while simultaneously articulating exclusion ("that although this museum is in England, they've got our stuff as well"). Importantly, she specified significant cultural knowledge that articulates belonging in this group, while also directing the conversation through relevant knowledge and everyday practices, revealing a relaxed and "dialogic space (where learning and communication is two-way)" (Tolia-Kelly 2007, 4). The fluidity, or 'liminality' (Tolia-Kelly 2007), between researchers and participants that these spaces afford results in opportunities for participants "to introduce terms and vocabularies that are more relevant to their experiences" (4–5). The significance of these religious practices as vocabularies of belonging and community can then be integrated into later conversations to ensure meaningful discussions in terms that are useful for participants. For instance, the significance of Hindu festivals and rituals in articulating community in Loughborough's Gujarati community became an important question in follow-up interviews. It could be said that these activities, both in their broader co-design and in the instances of fluidity during such interactions in the sessions, signal "moments of empowerment" (Ross 2017). These moments arise in interactions between the researcher and the participant and exist when spaces for equal dialogue between research and participant are built into the research process (Ross 2017, para 22). Importantly, these 'moments' acknowledge the limits of creating equal power relations, but they also show the possibilities of collaboration in the research process when working with creative methodologies.

Photography, locality and belonging: creative outputs

Creative methods also have potential within and beyond the research encounter through the dissemination of collaborative research outputs. Such outputs, we argue, create opportunities for community ownership of the project activities and have the potential to challenge disenfranchisement and (re)claim spaces of belonging (see Zahir 2003). For instance, we worked with a Bangladeshi women's ESOL group on the project on 'photography, locality and belonging' in London. Most of the group – aged between their twenties and their sixties – had arrived in London from Bangladesh, apart from three of them who had lived in Italy for some years before relocating. All members of the group migrated to join their husbands or families, their social life mostly revolved around family and their participation in the public life of the area is very limited. As a particularly hard-to-reach group, it would have been very difficult to reach out to them had we not been approached by their teacher, a local activist, who saw this collaboration as an opportunity to expand the range of activities for the group. The project was co-designed as an opportunity for British Bangladeshi women living in the area to narrate their neighbourhood through their own eyes while providing us the opportunity to discuss our research themes. Each participant was assigned a disposable camera and we took a guided walk around the neighbourhood, during which participants took photographs of places that they deemed significant for their own personal history. Walking around the area of Whitechapel and sharing ideas and comments with other participants triggered specific memories which led to participants deciding upon what places were more relevant for them. Some of the most photographed places included the East London mosque, the Ideastore library, Whitechapel market, the statue of Sheikh Mujibur Rahman and the Zakat centre in Whitechapel. The photographs were then used as a starting point to discuss migration from Bangladesh, their life in London and their engagement with the local community. As photographs offer a visual aid that facilitate mnemonic processes (Pickering and Keightley 2013), the photographs made it easier to draw connections between places and personal stories of migration, life in the neighbourhood and questions of identity and belonging. While almost all participants initially expressed their surprise at being asked about their own memories and experiences (instead of their husbands'), they found that they had a lot to say through the process and agreed to integrate their photographs, and reflections, into the exhibition "Aldgate East My Neighbourhood" at Ideastore Whitechapel, curated by Four Corners and Swadhinata Trust (see Figure 6.4).

The inclusion of the group's photographs and reflections in the exhibition represented a way for the group to make an intervention in a public debate (e.g., on the diverse character of communities in East London) from which they are mostly excluded. Moreover, the creation and inclusion of these outputs in a public space reflected a way to claim back space from a position of marginalisation. As Saskia Warren (2016, 801) notes, ethnic exclusions and inclusions can form part of the very construction of public space, along with other markers of class, gender, race,

FIGURE 6.4 Aldgate East My Neighbourhood exhibition, Idea Store Whitechapel, London. Participants visiting the exhibition. December 2018

religion, sexuality and age. Indeed, as Puwar (2004, 8) points out, social/public spaces "are not blank and open for any body to occupy. . . . There is a connection between bodies and space, which is built, repeated and contested over time". During our walks and discussions, the participants of our workshops clearly expressed that there are urban spaces they feel entitled to occupy, but at the same time there are others they feel excluded from; hence, when entering spaces of exclusion, they are akin to what Nirmal Puwar (2004, 8) calls "space invaders" – people whose bodies are marked out as "trespassers", where others are deemed as having the right to belong. Through the development of outputs that are exhibited in public arts and/or heritage spaces, community groups, particularly those who are most marginalised, can claim back new and old spaces and self-produce and self-represent diverse cultural identities (Tolia-Kelly 2007, 11–12; Zahir 2003, 207). The outputs of the creative activities themselves open opportunities to challenge disenfranchisement and (re)claim space both within and beyond the research process.

Creative methods and community activities: challenges

While creative, participatory methods have myriad advantages, such work presents numerous challenges. Before concluding the chapter, we briefly highlight several of the issues that have arisen using this approach.

The group discussions that arise from community activities are one of our primary modes of data collection in our creative methodological approach and the dynamics of these group discussion is significant in numerous ways. In terms of researching memory, the group dynamic can trigger personal and cultural memories between individuals and provide a window into social processes of remembering in action. While we advocate this type of group interaction and data collection, group discussions also present challenges and additional issues. In practical terms, group discussions during creative activities can quickly become overlapping conversations among participants, in worst cases, rendering our research recording unusable. While interventions in the discussion can solve this problem, too many attempts to gain more control of "which voices talk when, how much, and in what order" (Feld 2012, 241) can become counterproductive to methodologies that aim to be more egalitarian. While John Kuo Wei Tchen (2006, 200) notes that multicultural, collaborative projects and person-to-person dialogic spaces need explicit nurturing and curating, we have to strike a balance in our intervention to simultaneously gain useable data, support the dynamic of the existing group and maintain an ethos of collaboration within the activity itself. More challenging in such discussions and participation overall is the playing out of internal group hierarchies and politics, in that individuals with more power in the community group become those voices that are heard. In this case, one-to-one fieldwork interviews become the site of privileging the voices of participants involved in the creative methods who may also be subject to inter-community hierarchies and power imbalances. More broadly, the multiple hierarchies and power balances within and between community groups reminds us the Other ought not to be essentialised as powerless in the essential process of researchers, especially those that come into marginalised communities as cultural outsiders, to identify their privilege (Enria 2016, 326). Acknowledging the multiple layers of power and hierarchy at play thus becomes a significant part of the reflexive process in our creative approach.

While working with community groups and within their dynamics and internal politics, we also become embedded in the navigation of community politics with the potential risk of our involvement aggravating pre-existing issues, hierarchies and/or sensitivities within the group. Our positionality within communities in diverse areas becomes all the more complex, particularly in a project examining intercommunal relations. With numerous overlapping community groups and networks of communication between and within such groups, spending comparatively equal amounts of time with different community groups is a practical challenge. With more collaboration and immersion with community groups comes more complicated interactions and multiple layers of responsibility (see Enria 2016, 325). The concern of the impact of the researcher's effect on the field is not a new one in ethnographic research; however, in the pursuit of equality, transparency and collaboration, such participation-centred interventions move further towards the spotlight. Reflexivity in both our research practices and our interactions with different community groups becomes all the more significant in such a research process.

Finally, while creative methods and dialogic, empathetic spaces for discussion pave the way for multiple voices and claims to knowledge, they do not provide access to all personal memories and histories. In researching memories of Partition and migration, memories and experiences of communal violence, racism and discrimination may also be actively left out by participants. During creative activities and interviews, first-generation participants, in particular, have subverted experiences of racism, with accounts of these experiences only surfacing in rare, off-the-record settings. A strategic process of forgetting may be taking place, and questions turn towards the ethics of 'knowing' or 'grasping' such accounts into our data. While such memories are significant for our study, we should also acknowledge there are certain aspects of participants' lives that we should not assume we have the right to know (Enria 2016, 325; Lather 2002, 213). Enria (2016, 325) suggests that the limitations of knowledge, even when collectively produced, may be a symbol of respect for the integrity of others' subjectivity rather than simply a barrier. The 'right to know' is a broader issue in social research, but the acknowledgement of subject-to-subject relationships between researchers and participants and respecting subjectivities in numerous ways is a significant reflexive step towards collaborative understandings of community and belonging. Despite these challenges, by using creative methods we both acknowledge the political, social and ethical importance of moving towards more democratic means of research and aim to meaningfully integrate multiple voices and claims to knowledge in our research.

Conclusion

Through creative, participatory, ethnographic methodologies, we are starting to collaboratively uncover layers of memory and belonging across different community groups in order to understand the consequences of Partition in South Asian Britain today. Combining tried and tested ethnographic methods with creative activities creates opportunities for multiple ways of knowing and experiencing the past, enabling a deeper and richer understanding of the nature and role of cultural memories of Partition and migration in the articulation of community and belonging in contemporary South Asian Britain. The integration of creative practices in our research also brings us towards a more nuanced understanding of everyday media in the communication of difficult pasts.

In this chapter, we have argued that creative participatory methods provide solutions to methodological challenges in researching memory because they provide opportunities for participants to approach feelings of belonging, discrimination, isolation and marginalisation in oblique ways and provide safe expressive spaces for their articulation. They are moments, not for uncovering hidden meaning but of shared knowledge production in which the making of meaning is analogously intertwined through the cultural forms of making, from cooking to sewing. In contexts of migration, creative participatory methodologies are also profoundly important in facilitating inclusion, including the creative regeneration of identities, communities and subjectivities, and are integral to cultural citizenship (understood as "the right to presence and visibility, not marginalization; the right to dignity and maintenance of lifestyle, not assimilation to the dominant culture; and the right to dignifying representation, not stigmatization") (O'Neill 2013, 53). Such an argument for the use of participatory arts methodologies in contexts of migration, diaspora and in experiences of exclusion, discrimination and racism, as a means of harbouring cultural citizenship, also reflects the ethos of co-production and the shift in presence, visibility and representation of marginalised communities in the research process.

Longitudinal research design allows for ongoing negotiations with community groups and for cyclic processes of reflexivity and responsiveness in the methodological approach and research design, integrating co-produced and multi-perspective experiences and meanings of community and belonging and moving towards a multi-voiced, more egalitarian approach. Constant problematisation and reflexivity in assessing power relationships; being self-critical about assumptions, judgements, expectations and self-perceptions; as well as embracing discomforts and challenges (Enria 2016, 325; Tolia-Kelly 2007, 4–5) is a first step towards confronting and shifting the unequal power structures in research. To facilitate such an approach is not an easy task and holds numerous challenges and limitations, but creative community activities, as well as long-term engagement and immersion in the field, provide a path towards developing the necessary relationships, trust and rapport that may result in more egalitarian knowledge creation. While such research, MMPI included, serves the researcher's agendas, by starting with participation-centred research, creative methods and a reflexive and responsive approach are vehicles for researching memory in a more egalitarian way.

Notes

1 We refer to first generation here in relation to migration to the UK rather than first-hand experience of Partition. The interviewee did not directly experience Partition.
2 For a comprehensive overview of research design and methodology for empirical research in cultural memory, see Kuhn (2002).
3 Notwithstanding the 'new history' of Partition (see Butalia 2000; Virdee 2013, etc.).

References

Abarca, M.E. and Colby, J.R. (2016) Food memories seasoning the narratives of our lives. *Food and Foodways* 24(1–2): 1–8.
Ahmed, S. (1999) Home and away: Narratives of migration and estrangement. *International Journal of Cultural Studies* 2(3): 329–347.
Bakrania, F.P. (2013) *Bhangra and Asian Underground: South Asian Music and the Politics of Belonging in Britain.* Durham, NC: Duke University Press.
Barnes, A.J. and Kraamer, M. (2015) Japanese Saris: Dress, globalisation and multiple migrants. *Textile History* 46(2): 169–188.
Baumann, G. (1996) *Contesting Culture: Discourses of Identity in Multi-ethnic London.* Cambridge: Cambridge University Press.
Beebeejaun, Y., Durose, C., Rees, J., Richardson, J. and Richardson, L. (2013) 'Beyond text': Exploring ethos and method in co-producing research with communities. *Community Development Journal* 49(1): 37–53.
Bissell, S., Manderson, L. and Allotey, P. (2000). In focus: Film, focus groups and working children in Bangladesh. *Visual Anthropology* 13(2): 169–183.
Brah, A. (1996). *Cartographies of diaspora. Contesting identities.* London: Routledge.
Butalia, U. (2000) *The Other Side of Silence: Voice from the Partition of India.* Durham, NC: Duke University Press.

Carolan, M.S. (2011) *Embodied Food Politics*. Surrey: Ashgate.

NOMIS (2011). *Charnwood Local Authority: Local Area Report (ONS – 2011 Census (KS201EW)*. https://www.nomisweb.co.uk/reports/localarea?compare=E07000130#s ection_6 (accessed 15th September 2021).

Chakraborty, K. (2009) 'The Good Muslim Girl': Conducting qualitative participatory research to understand the lives of young Muslim women in the *bustees* of Kolkata. *Children's Geographies* 7(4): 421–434.

Choo, S. (2004) Eating *Satay Babi*: Sensory perception of transnational movement. *Journal of Intercultural Studies* 25(3): 203–213.

Clifford, J. (1986) Introduction: Partial truths. In *Writing Culture: The Poetics and Politics of Ethnography*, edited by James Clifford and George E. Marcus, 1–26. Berkeley: University of California Press.

Das, V. (1998) Specificities: Official narratives, rumour, and the social production of hate. *Social Identities* 4(1): 109–130.

Dudrah, R. (2006) *Bollywood. Sociology Goes to the Movies*. London: Sage.

Enria, L. (2016) Co-producing knowledge through participatory theatre: Reflections on ethnography, empathy and power. *Qualitative Research* 16(3): 319–329.

Fabian, J. (2012) Cultural anthropology and the question of knowledge. *Journal of the Royal Anthropological Institute (N.S.)* 18: 439–453.

Feld, S. (2012) *Sound and Sentiment: Birds, Weeping, Poetics, and Song in Kaluli Expression*. Durham, NC: Duke University Press.

Feuchtwang, S. (2000) Reinscriptions: Commemoration, restoration and the interpersonal transmission of histories and memories under modern states in Asia and Europe. In *Memory and Method*, edited by Susannah Radstone, 59–77. Oxford: Berg.

Foucault, M. (1981) *The History of Sexuality. Vol. 1*. Harmondsworth: Penguin.

Gunaratnam, Y. (2003) *Researching 'Race' and Ethnicity: Methods, Knowledge and Power*. London: Sage Publications.

Harper, D. (2002) Talking about pictures: A case for photo elicitation. *Visual Studies* 17(1): 13–26.

Hoque, A. (2015) *British-Islamic Identity: Third Generation Bangladeshis from East London*. London: Institute of Education Press.

Kabir, A.J. (2009) Hieroglyphs and broken links: Remediated script and partitions effects in Pakistan. *Cultural and Social History* 6(4): 485–506.

Kearney, A. (2013) Ethnicity and memory. In *Research Methods for Memory Studies*, edited by Emily Keightley and Michael Pickering, 132–148. Edinburgh: Edinburgh University Press.

Keightley, E. and Pickering, M., eds. (2013) *Research Methods for Memory Studies*. Edinburgh: Edinburgh University Press.

King, N. (1997) Autobiography as cultural memory: Three case studies. *New Formations* 30: 50–62.

Kuhn, A. (2002) *An Everyday Magic: Cinema and Cultural Memory*. London: I. B. Taris Publishers.

Lather, P. (2002) Postbook: Working the ruins of feminist ethnography. *Signs: A Journal of Women in Culture and Society* 27(1): 199–227.

Lerpiniere, C. (2013) One wedding, two cultures, four outfits: The phenomenological exploration of fashion and textiles. *The Journal of Textile Design, Research and Practice* 1(1): 27–41.

Mankekar, P. (2015) *Unsettling India: Affect, Temporality, Transnationality*. Durham, NC: Duke University Press.

McLoughlin, S. (2013) Research paper WBAC 012 discrepant representations of Multi-Asian Leicester: Institutional discourse and everyday life in the 'model' multicultural city. *From Diasporas to Multi-Locality: Writing British Asian Cities AHRC Research Network*: 1–53.

O'Neill, M. (2013) Women, art, migration and diaspora: The turn to art in the social sciences and the 'new' sociology of art? In *Women, the Arts and Globalization: Eccentric Experience*, edited by Marsha Meskimmon and Dorothy Rowe, 44–66. Manchester: Manchester University Press.

Passerini, L. (1983) Memory. *History Workshop Journal* 15(1): 195–196.

Pickering, M. and Keightley, E. (2013) Vernacular remembering. In *Research Methods for Memory Studies*, edited by Emily Keightley and Michael Pickering, 97–112. Edinburgh: Edinburgh University Press.

Puwar, N. (2007) Social cinema scenes. *Space and Culture* 10(2): 253–270.

Puwar, N. (2004) *Space Invaders: Race, Gender and Bodies Out of Place*. Oxford: Berg.

Radstone, S., ed. (2000) *Memory and Method*. Oxford: Berg.

Ross, K. (2017) Making empowerment choices: How methodology matters for empowering research participants. *Forum Qualitative Sozialforschung/Forum: Qualitative Social Research* 18(3): Art. 12. https://doi.org/10.17169/fqs-18.3.2791

Said, E.W. (2003) *Orientalism*. London: Penguin.

Sardar, Z. (1998) Dilip Kumar made me do it. In *The Secret Politics of Our Desires: Innocence, Culpability and Indian Popular Cinema*, edited by Ashis Nandy, 19–91. London: Zed Books.

Shukla, S. (2001) Locations for South Asian diasporas. *Annual Review of Anthropology* 30: 551–572.

Tchen, J. K. W. (2006) On forming dialogical-analytic collaborations: Curating spaces within/between universities and communities. In *Identity Politics Reconsidered,* edited by Linda Martín Alcoff, Michael Hames-García, Satya P. Mohanty and Paula M. L. Moya, 193–208. New York: Palgrave Macmillian.

Tolia-Kelly, D.P. (2004) Materializing post-colonial geographies: Examining the textural landscape of migration in the South Asian home. *Geoforum* 35(6): 675–688.

Tolia-Kelly, D.P. (2007) Participatory art: Capturing spatial vocabularies in a collaborative visual methodology with Melanie Carvalho and South Asian Women in London, UK. In *Participatory Action Research Approaches and Methods: Connecting People, Participation and Place,* edited by Sara Kindon, Rachel Pain and Mike Kesby, 132–140. New York: Routledge.

Tower Hamlets Council. (2013). Ethnicity in tower Hamlets: Analysis of 2011 census data. *Research Briefing.* https://www.towerhamlets.gov.uk/Documents/Borough_statistics/Ward_profiles/Census-2011/RB-Census2011-Ethnicity-2013-01.pdf

Tower Hamlets Council. (2018). *Population Projections for Tower Hamlets.* www.towerhamlets.gov.uk/Documents/Borough_statistics/Population/Population_Projections_for_Tower_Hamlets.pdf

Tuhiwai Smith, L. (2012) *Decolonizing Methodologies: Research and Indigenous People*. London: Zed Books.

van der Vaart, G., van Hoven, B. and Huigen, Paulus P.P. (2018) Creative and arts-based research methods in academic research: Lessons from a participatory research project in the Netherlands. *FQS Forum: Qualitative Social Research* 19(2): 1–30.

Virdee, P. (2013) Remembering partition: Women, oral histories and the Partition of 1947. *Oral History* 41(2): 49–62.

Warren, S. (2016) Pluralising the walking interview: Researching (im)mobilities with Muslim women. *Social & Cultural Geography* 18(6): 786–807.

Williamson, A. and DeSouza, R. (2007). *Research with Communities: Grounded Perspectives on Engaging Communities in Research*. Auckland: Muddy Creek Press.

Yuval-Davis, N. (2006). Belonging and the politics of belonging. *Patterns of Prejudice* 40(3): 197–214.

Zahir, S. (2003) 'Changing views': Theory and practice in a Participatory Community Arts project. In *South Asian Women in the Diaspora*, edited by Nirmal Puwar and Parvati Raghuram, 201–213. Oxford: Berg.

7

DRAWING TOGETHER, THINKING APART

Reflecting on our use of visual participatory research methods

Divya Patel and Lauren McCarthy

Introduction

Recent developments in creative methodologies and methods have opened up opportunities to build understanding about human experience in social systems, including within qualitative organisational and management research. When joined together with feminist principles of encouraging voice, inclusion and reflexivity, and challenging power relations in research processes (Caretta and Riaño 2016; Gatenby and Humphries 2000; Reinharz 1992), creative methods can become vehicles for empowering research. In this chapter, we outline a case study into corporate social responsibility (CSR) in Gujarat, India, that employed visual participatory research (VPR) methods applied alongside other traditional research methods of in–depth semi-structured interviews. We draw on our experiences to argue that VPR methods, used creatively and flexibly, can illustrate organisational and social phenomena from different standpoints, particularly those of women, but also of men, living and working within patriarchal, caste-based systems. Drawing, alongside verbal reflection, can play an important role in beginning the challenging, yet ethically imperative process of empowering research (Davis 2012). However, these processes are not easy. We share our own experiences and learning in facilitating free-hand participant-led drawings through three reflective lenses: first, our 'being', as researchers; second, Divya's 'being' as an 'insider' to Indian culture and Lauren's 'being' as an 'outsider'; and third, by 'being' women in the data generation process.

Drawing as a participatory visual method

Visual methods have had something of a renaissance in business and management studies. There are now several books (Bell, Warren and Schroeder 2013; Höllerer et al. 2019) and review articles (Davison, McLean and Warren 2012), as well as different empirical approaches to generating and analysing visual data, including

DOI: 10.4324/9780429352492-7

the use of drawing methods. Drawing as a visual method can take the form of participants creating diagrams (Bagnoli 2009; McCarthy and Muthuri 2018) or more free-hand sketches (Clarke and Holt 2019; Vince and Broussine 1996; Zuboff 1988) in response to questions about their experiences. By focusing on drawing as one form of visual participatory research (VPR) (McCarthy and Muthuri 2018), we aim to emphasise the empowering research potential of both the visual and the participatory. Indeed, the two cannot be separated since not only are visual data created but interaction, dialogue and increased participation are central to the approach (Stiles 2013). The aim here is to find new ways of challenging the power imbalance between the researcher and the researched (Gallagher 2008), which has both ethical and practical imperatives (Pauwels 2015). From a 'scientific' knowledge-generating perspective, drawing puts participants in a position of expertise, as they draw and narrate their images (Pauwels 2015), sometimes enabling an openness and richness that other methodological techniques may not provide (Literat 2013).

Whilst the practical research benefits of VPR methods are easier to defend, their actual empowering potential, and thus ethical position, remains hotly contested (Oliveira 2016). Empowering research is purposively aimed at leaving participants in a better position than before they took part in the study, for example, through the catharsis of sharing experience or in aiding 'voice' to lobby for changes to people's lives and communities (Davis 2012; Ross 2017). "Ensuring fewer exploitative and hierarchical research relationships" (Caretta and Riaño 2016, 263) through participant and researcher "involvement, activism and social critique for the purpose of liberatory change" (Gatenby and Humphries 2000, 89) is a central tenet of feminist participatory research. Thus, feminist research shares epistemological roots with 'empowering research' methodologies which focus on interpretation, dialogue, voice and power (Davis 2012). Feminist research ensures that women's and other oppressed people's voices are centred (Maguire 1996). Research is not a politically neutral activity and thus both feminist and empowering research approaches share a responsibility to 'do no harm' at a minimum and to promote individual or social change at best (Caretta and Riaño 2016; Maguire 1996; Reinharz 1992). Poststructuralist and postcolonial feminist (e.g. Butler 1990; Spivak 1988) scholars, however, have pointed out that the idea that an incomer could know what constitutes 'good' change is problematic (Varga-Dobai 2012). Notwithstanding this, feminist research seeks to enable reflection and space in which the seeds of change – defined by participants themselves – are sown.

Yet there is no guarantee that drawing as a VPR approach can achieve these aims. This is in part due to the varying degree of participation that such approaches involve (Chalfen 2011), the challenges of following a feminist perspective in what remains a strongly positivist research culture within the social sciences (Gatenby and Humphries 2000) and the limits of genuine collaboration within increasingly neoliberalised research spaces (Bell, Kothiyal and Willmott 2017). For example, limited time and money may push researchers towards an easier research design where drawings are closely curated, and participant creativity is limited. Our own situatedness as researchers, with our racialised, gendered and classed baggage, plays into the empowering research process. In the remainder of this chapter we recount the ways in which we have used drawing as a VPR method in our CSR research, why we do so and the questions that this raises for us related to participation, voice and empowerment. We begin by introducing a case study into corporate social responsibility in Gujarat, India, that employed visual participatory research methods, which we use to illustrate our arguments.

Understanding meanings and relationships in business

How do married couples who are small-business partners understand and enact CSR? What are the dynamics of their work and home lives? How do systems of gender, race, caste and class interact with these relationships? These were the motivating questions undertaken by the first-named author in Gujarat, India, as part of her PhD. India presents a context where many couples in a business deal with intertwinement of family and business activities (Matzek, Gudmunson and Danes 2010), and share ownership, commitment and responsibility of the business (Fitzgerald and Muske 2002) for economic (Fletcher 2010) and social reasons. For example, in the patriarchal context of Gujarat, women and men often claim that it is better if married women join their husband's business rather than work elsewhere as this enables the maintenance of women's primary caregiving role alongside contributing to the family business (Budhwar, Saini and Bhatnagar 2005). Women are also afforded less decision-making whilst taking on more responsibility but largely remaining unrecognised and unrewarded (Heinonen and Stenholm 2011).

Within such a context, at first, in-depth interviews appeared an appropriate method to discuss how jointly lived experiences between the couple shaped their understanding of CSR. However, this raised questions related to the possibility of egalitarian participation by both partners in the data generation process, namely: during the interview, will one partner dominate another? How will the researcher respond to these inequalities? Will the research be trustworthy if one voice is suppressed in the conversation? How can perspectives be brought forward from both partners? To tackle these issues, partners were interviewed separately with each participant asked the same questions. Separate interviews allowed participation and addressed the problem of unequal distribution of power between the couple. However, questions about the researcher's own position of power as the one asking questions and leading the conversation remained a challenge. Since the study was interested in 'understandings' of CSR, there was also a need to find techniques for helping participants to articulate this. It is here that the idea of using drawing as a VPR method in addition to the interview arose.

Research design

The research design helped tackle this challenge and enabled participants to take control of their own expressions by creating drawings. In addition to face-to-face semi-structured interviews, the study used participant-led drawing with

54 participants (27 heterosexual partners) who ran a small to medium business together in Gujarat, India. Following the interviews in which the participants reflected upon what being socially responsible in business meant to them, both partners were brought together and were each given an A4 piece of plain white paper and a pen and asked to create a free-hand drawing. The instruction given was: 'draw your business doing socially responsible activities'. Conceptualising the business as a social actor enabled a broader consideration of who gets to decide what is defined as CSR, and who are included as stakeholders. If participants had been asked to 'draw yourself doing socially responsible activities', then the relational aspects of SME-related business and CSR may have been left out (Spence 2016), furthering the occluded nature of women's roles and contributions in the partnership. Furthermore 'doing' reflects a process which may have evolved or is altered due to historically situated ideas, embedded in cultural values and practices (Bell, Bryman and Harley 2018). Thus, when participants were drawing, these socio-culturally constructed influences were reflected in the visual image created.

Galvaan (2007, 156) suggests 'going slowly, taking time' in the drawing activity. Participants need enough time to visualise and to draw, as making the drawing is contingent on a process of reflection and of finding a way to express this pictorially. Participants may feel overwhelmed thinking about how they can put their words in an image form, but drawing being a form of meditating therapy (Rees 1998; Silver 2001), if given sufficient time, enables reflection on feelings, recollection of experiences, and making connections. To aid the participant drawing process, sample images were provided to inspire participants: objects such as a tree (nature, environment), a stick man/woman (since the human figure is difficult to draw), a plastic waste bottle (plastic usage) and a group of people (to indicate employees, family or community members). It was emphasised that these were just suggestions – not proscriptions.

Participants' drawings included: images representing their own role in the business, and those they share with their business/life partner; images symbolising their business in society; and connections and relationships to stakeholders, for example, represented by hearts, or arrows, or holding hands. Others drew connections and sequences through rings and layered pyramids, reflecting their CSR meaning-making process. They indicated priorities by circling the person or object, dotting their pen hard on a symbol while talking about it, making the object bigger compared to other objects or covering a larger part of the paper. The openness of the initial instruction intended that participants could include all the stakeholders and items they could think of, literally imagining and imaging their universe and their responsible role towards each object or the person within and around it. Furthermore, drawings enabled participants, especially women participants, to voice their situations even more than what they had shared in the interview, reflecting prevalent Indian societal norms. Examples of this are discussed further in the chapter.

Participants' explanations of the images they created were audio-recorded (with their consent) and observations, thoughts and reflections on the process as well as a photograph of the image itself were recorded in the researcher's diary.

Analysis

Despite methodological advancements, available analytical techniques to systematically analyse images are scarce (Drew and Guillemin 2014). Analysis of visual data can be a complex process as there are issues concerning multiple interpretations (Shortt and Warren 2019). Verbal and textual data such as interviews can also be interpreted in multiple ways; however, visual data is particularly prone to different interpretations as they embody contextual socio-cultural meanings which may require further clarity by the respondents (Prosser and Loxley 2007). This requires a keen understanding of the socio-cultural context guiding human experiences depicted in the images. In the absence of this, it can become especially challenging for someone outside the research interaction to articulate the meaning of images created, except for guessing certain generic features. Where there are pre-agreed or mutually understood symbols or diagrams, interpretation may be more systematic (e.g., McCarthy and Muthuri 2018), sometimes even carried out using content analysis (Bell 2001). However, free-hand drawings are more easily analysed when there is a combination of visual and verbal qualitative analysis of the drawing and the discussions about the drawing (Guillemin 2004) to gather a contextually grounded interpretation of the drawings (Meyer, Höllerer, Jancsary and Van Leeuwen 2013).

In the CSR case study, a Grounded Visual Pattern Analysis (GVPA) approach (Shortt and Warren 2019) was used. Adopting a recognised analytical framework which had been developed by leading researchers in the field intended not only to help make analysis more systematic but also to gain more legitimacy for a method that is still little-known within business and management studies. It is something that felt even more important given that this was PhD fieldwork which we hoped to publish from one day. GVPA has been used to analyse participants' photographs by analysing the discursive meanings that are attached to created objects and patterns, as well as influences of the social structures which embody these meanings (Meyer et al. 2013; Shortt and Warren 2019). This method enabled a systematic approach to the analysis of participant-created images, aiding the credibility of the research data.

Analysis of the drawings was broken into two phases. The first phase focused on participant meanings of their drawings in which the drawing as well as the explanation transcript was uploaded on NVivo 12 and further categorised and developed into codes and themes (Boyatzis 1998). These themes were also matched with existing first interview themes, generating emergent themes. In the second phase, analysis concentrated on understanding the social setting and reconstructing underlying meanings drawn in the images (Meyer et al. 2013). This involved the researcher's own interpretation of the drawing by relating the images to the key terms of the research and noting any emerging patterns, sequences, layers, metaphors, unsaid things which the participant may not be able to describe in words. This meant being particularly alert for symbols and images representing socio-religious influences (e.g. a cow to represent Hinduism), and social or structural arrangements reflecting traditional mindsets (e.g. the depiction of particular clothing).

Following the GVPA approach, images were ordered and grouped as per themes (Pink 2007; Shortt and Warren 2019). When considering these drawings in relation to the socio-cultural, religious, familial and societal norms within the Indian context, this helped to begin theorisation on partners' 'doing' of CSR. Grouping also enabled systematic analysis of images across participants, as well as deriving a comparative view of male and female participants.

Analysing drawings through GVPA enabled the inclusion of participant's voice in the data analysis process. Each participant's explanation of their drawing was included in the analysis, and therefore their own initial interpretation of their own drawing guided the visual data analysis. They could voice what may have been wrongly interpreted, and in this sense, VPR methods get closer to empowering research in that they address the gap between the researched, as spoken for, and the researcher as the speaker. Figure 7.1, a drawing by Shalini (all names are pseudonyms), provides an example of this. She draws four human figures in a line and explains that they symbolise equality. She clarifies she has intentionally not given gender-specific features like long hair or moustache to these figures as she feels everyone is the same. She asserts that she and her husband are equal and that their daughter is equal to their son. In her conceptualisation of CSR, she includes gender inequality as a pervading social issue which she consciously addresses in their family business as well as home practices. She shares adamantly convincing her

FIGURE 7.1 Drawing by Shalini

family members for her daughter's higher education abroad instead of the popularly preferred male child education.

In addition to uncovering gender inequalities experienced by Shalini, her drawing also opens up the conversation into her own experiences of being ousted and neglected by her spouse's family because of her caste. Without the drawing, the researcher's interpretation could have missed what are deeply felt, learned experiences of gender and caste in relation to running a business as a married couple. Further, the drawing process enabled her to share and reflect on how sad and hurt she felt when neglected for years, in what appeared to be a cathartic experience for her.

Having laid out the more practical research-based justifications for using VPR methods (Pauwels 2015), we now turn to more ethical considerations.

Reflecting on visual participatory research

We now reflect on our experiences of using VPR from a feminist perspective. To do this, we discuss our experiences of using VPR as researchers, as insider/outsiders and as women. We each let our own voices, as embodied individuals conducting our research, speak freely here.

Reflecting as researchers

Divya: I firmly believe that the research design involving VPR methods further enabled participants' voice and participation to take control of their own expressions by creating drawings; however, as a researcher, there was a need for constant reflective analysis of the research process. With pressures to conform to the demands of systematic research, tensions over credibility and purity of data, ethical dilemmas concerning participant consent, confidentiality and anonymity were persistent challenges (Bhattacharya 2007) demanding different ways that I address, or at least, reflect on this.

In the pilot data generation exercise, I interviewed the husband and wife jointly and asked them to co-create a drawing of their CSR understanding on a shared paper. The idea behind this step was to see how the couple worked together and mutually conceived/perceived ideas about CSR. However, during the drawing activity, only the female participant drew whilst describing the objects in her image in connection to their business's CSR activities. The male participant added to the discussion verbally but drew nothing. When one partner 'led' the drawing, this not only created problems of participation between participants but also raised questions of power between the couple. As a researcher wanting to gather rich data and adhering to the adopted empowering methodological approach, I wanted to make sure that an ambience of participation was created, and all participants were provided an opportunity to express their views on CSR. This was especially important given the focus on gender. Another challenge I faced during the pilot exercise was participants were not drawing as such but writing words, and in doing so, some resisted further discussion on the topic, wanting to wrap up the activity speedily.

Thus, to address this issue of participation and power, during the data generation in India, a separate sheet of paper was given to each participant, and they were asked to avoid using words. With reluctant participants, some different suggestions of how to start the drawing were offered, and the spouse's eagerness motivated the other to participate. At times, even a light competitive spirit between the couple, of presenting the image in best possible shapes and ways, encouraged non-enthusiasts. Some encouragement from my side and different ways of asking them to draw inspired them to accept that the task could be useful.

This activity worked well but observing the process of how couples mutually co-created a shared conceptualisation of CSR as a researcher and generating this process as 'data' was lost. However, it was essential that each individual participant was provided an opportunity to share and voice their thoughts, and thus this way, individual conceptualisations of CSR emerged. In line with the adopted social constructionist epistemological position in this research, the individual depictions of CSR allowed the surfacing of subjective identities and their respective personal characteristics, which influenced how they made meaning out of CSR.

Going through the pilot process of conducting the drawing activity, I realised that participants, when confronted with drawing, were not always as willing and as enthusiastic as I was. It was challenging for me to convince participants to take part in the drawing exercise. For example, I could not include insights of some participants in the analysis process, as after the interview was conducted, due to constraints or personal reasons, they could not participate in the drawing exercise. I could not include these participants in the analysis as the data was incomplete and not in line with the adopted data generation method of interviewing and creating drawings – challenging the ideal of systematic, robust data collection. Although I adopted visual methods to enable participants to fully express themselves, constraints of regulations, time and mental space had to be faced throughout the data generation process.

On the other hand, I was highly aware of my position as a UK university researcher. I observed that the power dynamics between the participants and me were much more visible and obvious in the interview sessions than in the drawing exercise. Although semi-structured in nature, the interviews took the course of a conversation. However, there was still a sense of my power as the one guiding/ leading the conversation. Whereas during the drawing exercise, the power seemed to shift to the participants, and my role took a backseat in the image interpretation process. Yet I was conscious of my physical presence whilst they drew as I did not want to rush them nor wanted to make them feel as if I was going to evaluate their drawings.

Several such challenges and questions lasted within me throughout the data collection and analysis process when working with VPR methods: would the participants give me additional time after they had already spent time with me in the interview? How will I analyse drawings? These more methodological concerns were joined by ethical ones. I constantly faced ethical dilemmas whilst protecting participant identity and confidentiality in dealing with visual data. Questions that bothered me were: can I show participant drawings to other PhD colleagues, friends and family members to gain their perspectives or interpretation on top of the one given by the participant? I also reflected upon relational concerns, including my researcher relationship with the couple as well as personal relationship within them. Questions that worried me were: will my questions result in any disagreement within the couple? How will I manage the disclosure of painful past stories? Some online ethical challenges that were new to me were: if I had scanned and saved an image as a pdf file, can I upload the participant drawing on online free file conversion websites to convert it into an image format? Who is seeing these files on the virtual cloud? Keeping in mind the ethical considerations of handling research data, I was always conscious as well as cautious about protecting the drawings and participant information. But what if something had escaped my attention?

Navigating these tensions of implementing empowering research, my experience as a researcher was that drawings helped participants to reflect, correct what had been said in the interview, and add new ideas, presenting a fuller account of their CSR understanding. As an empowering methodology, drawing also enabled participants to find their own voice within the mundane but hectic day-to-day practices in which they play multiple roles. For example, Kamla, during the interview, discussed the business's social responsibilities towards employees, clients and community members. She explained her familial responsibilities, acknowledging family members as essential stakeholders to herself and the business. Whilst creating the drawing, she reflected on her place and her own self within the business, family and CSR dynamics. In Figure 7.2, at times she feels she is ignored in work and home dynamics and explains that her own mental well-being and fitness is important (symbolised by the bicycle), to sustain all the following relationships that are drawn and her responsibilities towards them, drawn as flowing out from her.

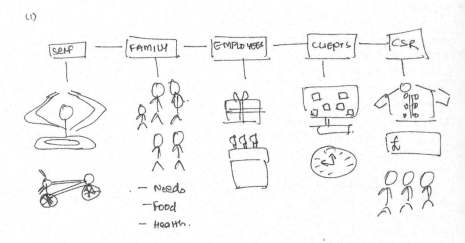

FIGURE 7.2 Drawing by Kamla

I found that drawing was also useful for the research as another means for participants to 'speak'. According to McIntosh (2010) certain experiences involving deeper emotions become difficult to explain and are better expressed through metaphors. Metaphors may therefore serve as a function for reflection, which allows participants to tackle powerful emotions strongly but in an indirect way that enables the expression of the feelings associated with those experiences (Broussine and Simpson 2008). For example, in his interview, Shankar described his frustrations regarding business-social relationships becoming a burden and how they are complicated in nature. He explained that he makes numerous efforts to sustain relationships with stakeholders, but thinks that he fails to deliver according to their expectations.

Shankar drew butterflies hovering over a lotus flower (Figure 7.3). He explained to me that the flower signified himself, the business and his spouse, which are blooming, emanating a fragrance and are in a happy space. He explained that the word 'rose' also has a personal reference to his relationship with his wife, Hansa. The butterflies

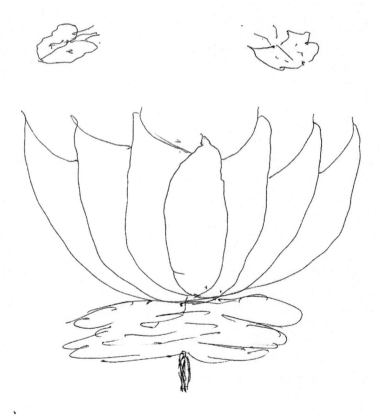

FIGURE 7.3 Drawing by Shankar

hovering over them symbolised social relationships which are bittersweet in nature but are close to them. There is reciprocity between the flower and the butterfly (business-stakeholder relationships), and that formed his meaning of CSR. The metaphor of butterflies helped Shankar to summarise his thoughts about CSR and social relationships that he had begun to think and express in the interview. Drawing skills or precision did not obstruct Shankar's thought process, and he was immersed in elaborating in-depth on this thought, I think, to generate clarity for himself as much as to me.

Lauren: The questions that go through my mind as a researcher as I use VPR methods are: will this be taken seriously? Will I be able to interpret the images when I get home? Will I be able to publish this as data? Innovative methods usually take more words on the page to explain; can excite and enrage reviewers seemingly equally; and, at times, can be frustratingly difficult to analyse – much less recount how that analysis was undertaken. Reading accounts of participatory visual methods in childhood studies, healthcare and development studies, I feel a sadness that my own field and employment doesn't enable me to take a longer time to engage with fieldwork. Chalfen's (2011) distinction between VPR 'studies' and 'projects', and how projects which engage long term in collaboration with local communities are more likely to be useful and empowering, cuts deep. My work feels stuck in a 'studies' category which means little meaningful, community-based engagement.

In some cases, introducing the pen and paper to the interview setting can make participants feel so uncomfortable, they are cringing and so am I. What am I doing, asking this senior manager to draw? Questions of ethics, particularly about the dynamics between myself and the 'researched' (which I reflect more on below) are central in my mind. A senior manager may feel more able to refuse to draw, but what about a cotton farmer? And if they do agree to draw, how can I ensure their anonymity is guaranteed if I publish the visual in an article?

On the other hand, as a researcher, I am in awe of the power of the visual to bring to life complex emotions, feelings and relationships that after weeks of interviews were not forthcoming. Take for example a recent research into a CSR partnership in India. Asking the manager of an NGO partner to describe the relationship with the corporation, he quickly drew a car, with a stick figure in the front, driving, and two smaller stick figures in the back seat. The corporation, at the front, was 'in charge, steering the show', whilst the two figures in the back were the NGOs, 'fighting like brother and sister'. Suddenly the dynamics of this particular partnership became much easier for me to understand.

And I always push back with reviewers – why is analysing a visual image any less objective (as if that is even possible) than analysing text? Form, meaning, subtext and interpretation are all at play in both forms of data.

Reflecting as insiders and outsiders

Divya: Being from India, my upbringing is rooted in Indian socio-cultural values, and thus cultural or social references made by the participants in their images resonated with me. I am aware of Indian cultural and religious traditions, day-to-day mannerisms of Indian homes and hierarchal etiquettes that are followed (e.g., the position of man in the family, a place for elders, the issue of space for in-laws from a female perspective). These interpretations were reflected in the participant's images as this was an inquiry into intertwined couple relationships. I believe being an 'insider' helped me to interpret this contextually embedded knowledge, which was already familiar to me due to my own lived local experiences (Brannick and Coghlan 2007). When such references were made, I felt able to ask relevant follow-up questions to let the participant settle deeper into those thoughts, feelings and emotions.

Similarly, in a culturally diverse country like India, social and cultural indicators take on more meaning. For example, within Hindu cultural tradition, a cow is considered one of the dearest living beings to God, and therefore people of Hindu religion, including Hitesh (Figure 7.4) respect cows and consider it to be their social responsibility to care for them. I understood what this symbol meant – would non-Indians do so? My knowing of the Indian context enabled Hitesh to further the conversation leading to more interaction over the research topic between us. This dialogue led to further reflexivity by the participant and perhaps empowered him to share and express meanings understood by both of us, in a moment of co-interpretation. Co-interpretation entails the intertwining of the participant's intended meaning with the researcher's understanding of the local socio-cultural influences informing the drawing (Banks 2001).

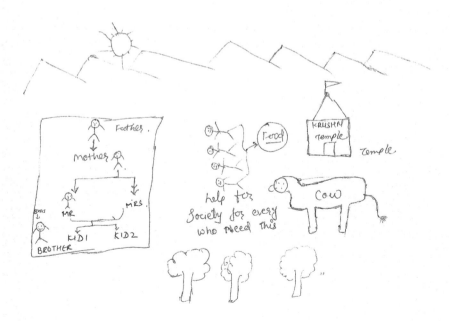

FIGURE 7.4 Drawing by Hitesh

Being conversant in local languages such as Hindi and Gujarati enabled me to encourage participants to express ideas in their everyday language. So, participants who could not speak English were not marginalised and missed but were integrated into the research. Some couples expressed their wish to speak in Hindi or Gujarati beforehand; others began the conversation in English but sometimes would use a rush of local words to express what they wanted to more fully. I could switch languages to understand better what participants wanted to convey. Participants also used Hindi/Gujarati words in their drawings to express themselves.

However, my being an Indian in no way suggests that my position is unproblematic or stable. Rather the dual role of being a researcher and an 'insider' comes with challenges of prejudice, the danger of presuming participant responses and undertaking a methodological course that restricts/blocks unknown perspectives. Continually reflecting on and analysing my role as an insider helped navigate the problems of presumed knowing and pre-understanding of participant responses (Brannick and Coghlan 2007). I would keep in mind questions such as: are my own Indian values influencing the participant responses and the analysis of data? How do I adapt to be ignorant of my Indian experiences and allow reframing of alternative perspectives? Taking time to be introspective, to expose any underlying assumptions (Argyris, Putnam and Smith 1985), and (re)think my own situations, alongside employing empowering methods such as drawings, helped tackle these challenges.

As an insider, I tried to be aware of my own presuppositions about the Indian context, which were constantly challenged throughout the data generation process. For instance, as an insider woman researcher, on the one hand, I was embracing changes in Indian men's perceptions of women but on the other hand was also frustrated by what I saw as the (re)creation of traditional gender roles, which ascribed power to masculinity and weakness to femininity. Figure 7.5 features a drawing

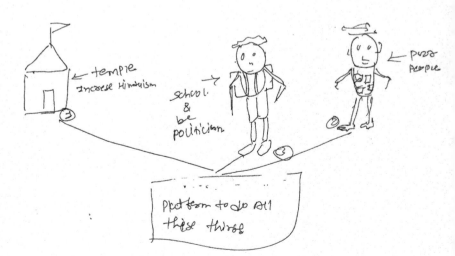

FIGURE 7.5 Drawing by Rohit

by a male participant, Rohit, which sheds light on my dilemma as an 'insider' researcher.

Rohit draws his primary CSR responsibilities as promoting Hinduism and serving the poor community – themes I had become familiar with. But then I was surprised when he went on to draw his daughter dressed like a boy in the centre. Rohit explained that as part of social reforms in women's lives, he would like to see women physically and mentally strong, 'like men' to tackle societal challenges. "I want to raise girls as boys", said Rohit. He wished to train their little daughter in martial arts and prepare her to become capable of protecting herself. I was taken aback in two ways, which challenged my positionality as an Indian. First, in that he had been openly supportive of women's rights. Having lived liberated yet subordinated roles to males in India myself (being a wife, being employed under a number of male bosses), I had faced the inherent patriarchy and frequently encountered situations where issues concerning women's empowerment are secondary for men. I was therefore surprised by Rohit's assertion of the need to promote women's empowerment. Rohit's commitment to women's empowerment was contradictory to my pre-understanding of how men in India perceive this pervading social issue. However, I also felt uncomfortable with Rohit's conceptualisation of gender, which clearly assigned weakness and subordination to femininity and ascribed power, autonomy and strength to masculinity.

Being an insider has strengths and weaknesses when it comes to using VPR methods. But had I not been from India, researching India beforehand in-depth would help in collecting data. But it does make me think, how much of it would have been enough?

Lauren: To answer Divya's question – it's never enough. You can read tens of books and talk to experts, but to research something like meanings, gender and CSR in a country or culture other than your own, as I do, means to be continually cast adrift as an outsider. From wearing the wrong kind of clothes to getting greetings or customs muddled, to constantly being a source of interest (at best) and harassment (at worst) in public spaces. Then there's the question of practicing VPR methods. For me, working alongside sensitive, intelligent research assistants has been the element that makes or breaks the research. I've had experiences where translators cut off participants talking, telling me, "She's not mentioning anything relevant". Others have tried to 'help' participants draw. This is difficult because the intention is well-meaning but the effects on the research can be disastrous. Further, I felt awkward and rude to be 'correcting' these behaviours. On the other hand, I have worked with research assistants who have painstakingly helped me translate interview guides into local languages, who have not only understood the research objectives but positively contributed to the design, and who have helped interpret drawings alongside my own 'outsider' readings.

On the other hand, doing ethnography or situated research that VPR methods demand is premised on the very notion of being an 'outsider' (Agar 1996). This means that everything is interesting, and my ignorance can be turned into useful inquisitiveness. Asking questions, making observations, pointing out inconsistencies – all of this is somewhat easier when you're not 'at home'. I have felt at times that my asking people to draw, as an outsider, was in some sense more palatable. I would obviously not understand, being a foreigner (in every way), how silly this request was. The activity is therefore begun with humour, intended to humour me! But this distance to the place I'm studying has become more of a concern as I continue doing this kind of research. It's not lost on me that all the countries in which I've researched – Sierra Leone, Tanzania, Ghana, India – were all exploited by and ruled over by the British. That relationship, of 'colonial researcher' visiting 'colonised people', cannot be wished away.

Another concern is the privilege and position I take as a White, British, woman in her mid-thirties with her own assumptions and biases around gender, race, ethnicity and so forth. The very topics of many of my research projects. As feminist researchers have reflected (Harding 1986; Mohanty 1991), we bring this baggage with us and it can be a painful process to realise that your feminism may not be the appropriate feminism for others. Hinson Shope (2006) shares field notes where she perceives South African women to be disempowered due to their deference to their husbands. She writes how her "Western feminist narrative about gender with an emphasis on contemplative individualism and critical self-analysis were being read into their gender narratives" (Hinson Shope 2006, 171). This chimes with an early realisation during my PhD that pathways to gender empowerment are not built by the outsider, in any sense of the term (McCarthy 2017). But being aware of the need to situate analysis in its context is not the same thing as being able to do this well (Opie 1992). This results in what I perceive to be a tension between wanting my research to contribute in meaningful ways – to organisational practice, to better working conditions for workers – and avoiding a neocolonial framing and analysis (Girei 2017).

I reflected, during one of my recent return trips to Gujarat, whether I was right to be carrying out this research and running focus groups. I was talking to women cotton farmers in very rural areas of Gujarat. What right did I have to be taking up these women's precious time? As a White British woman unable to speak Hindi or Gujarati, what was I doing here? Surely, I was hindering rather than helping the research, especially when the research assistant was doing such a great job. How can I use the findings I have to good effect, beyond another journal article? Is what I'm doing extractive (Spivak 1988), colonising knowledge from the global South and personally benefiting from its outputs in the North (Bell et al. 2017; Mama 2007)? And in which case, isn't what I am doing the opposite of empowering methodologies? But in stopping wouldn't I abdicate some of my political responsibility to use my skills to speak out against organisational wrongdoing (Alcoff 1992)?

I still have no answers to these questions, but I continue to ask them in the meantime.

Reflections as women

Divya: During my research, I could relate to women participants' stories. I felt I could see what the female participant was trying to draw and explain, as I have experienced strong patriarchal influences in India. For example, the struggle of being a wife juggling home and business, the insistence that home chores are not to be carried out by men and, frustratingly, that the final say on decisions is made by the 'man of the house'. My being a woman prompted me to relate to what female participants were saying, mostly because I have had similar experiences. For example, Riddhima (Figure 7.6) shares that she is torn apart by home and work. She verbally explained how she integrates elderly members of the society as her closest stakeholders in terms of CSR but carefully draws a longer arrow pulling her towards home, whilst writing the times she is expected to work in the shop. In the discussion, she further explained that she has the full responsibility for all home chores.

I think I would never have understood the way these patriarchal forces *feel* had I not gone through similar experiences. I could relate emotionally and physically to Riddhima's experiences, having lived similar gendered experiences. From a feminist research perspective, my own subjectivity may be interwoven in the analysis of the data; however, I did not intend to (re)create gender stereotypes based on my own experience but to carry out an inclusive analysis through reflexivity (Prosser and Loxley 2007). An inclusive account entails the meaningful collaboration of experiences of the participants with the researcher's views for social transformation (Byrne, Canavan and Millar 2009). It echoes strands of emancipation in consensus with participation and convergence of perspectives.

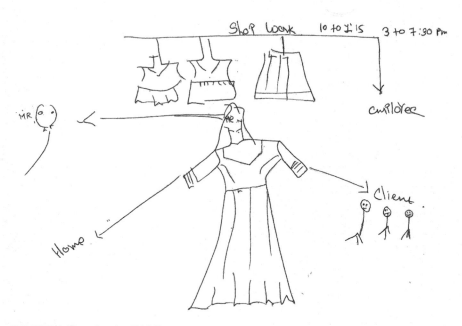

FIGURE 7.6 Drawing by Riddhima

VPR methods can help activate the agency of silenced voices (Eldén 2013). According to Warren (2005, 870), voice can be explained as 'the right or opportunity to express a choice or opinion' with images and speech. In highly stratified socio-cultural domains, such as Gujarat, certain stakeholders with lower incomes (watchmen, housemaids, drivers, slum communities), from 'lower' class groups (sweepers, maids), from religious groups other than their own, and women, maybe treated differently and denied a voice due to hierarchal discrimination, socio-religious norms and gendered structures. Employing interviews only as the data collection method may result in missing out the most vulnerable and their potential input into research. For example, the comparative view of Rashesh and Sangeeta's drawings (Figures 7.7 and 7.8) clearly indicates Sangeeta as expressing her voice, represented by her central role in attending to various stakeholders around her. She includes family members as key stakeholders by creating their larger images (father-in-law, husband, children) and includes business stakeholders in the small drawn business unit. Rashesh depicts his leading role in the business

FIGURE 7.7 Drawing by Sangeeta

FIGURE 7.8 Drawing by Rashesh

and includes business stakeholders. Drawing allowed Sangeeta to express how she is the focal point connecting all concerned members, and things would not function as they are without her. Creating the visual presented her with an opportunity to voice her situation.

Reflecting on my own experiences of being a woman particularly drew my attention to how men and women research participants conceptualised gender in CSR. This led me to tackle my positionality of being a woman whilst generating and analysing data through constant introspection and deconstructing my own gender biases to allow alternative perspectives to transpire.

Lauren: My experiences of using VPR methods are that the visual can be used to represent patriarchal norms, values and practices as much as any other form of method. Like Divya's experience in her pilot study, if men and women share paper, then men can dominate the drawing process, or refuse to engage altogether. If women are asked to draw in front of a mixed group of peers, they may decline and defer to a man in the group. So, VPR methods can facilitate 'voice' for marginalised groups of people, but careful planning needs to go into making this happen (Mayoux 2012). I have faced frustrations leading VPR groups, where gatekeepers such as NGO workers are dismissive of my experience, probably a combination of my perceived age and gender. Here is where I have battled for *my* voice in the research process.

Visual methods have been useful for making visible the invisible. In my research in Ghana, drawing helped capture economic tasks and re-orient their value in participants' eyes. For example, when speaking to female farmers, they spoke of themselves as 'helping' men on the cocoa farm. The diagram exercises (McCarthy and Muthuri 2018) that we ran enabled women and men to see that 'helping' consisted of carrying out nearly half of the cocoa farming tasks, including those such as fermenting and drying which contribute to better quality cocoa. Women were not 'just' helpers but skilled cocoa farmers. The exercises also captured the unpaid (and unrecognised) care work that women were contributing to the cocoa sector. As women circled the tasks which took up most of their time on their drawing, it became increasingly clear why more of them were not engaged in further economic or CSR-funded livelihood activity (McCarthy 2018).

Jackson (2012) makes the argument that 'voice' is often correlated with power, but that for many women in different parts of the world, 'voice' might be present in other, non-verbal ways. In this sense, visual methods can surface this, in ways such as those Divya mentions. The power of combining the visual with the verbal needs reiterating too, as women and men can draw on their sketches to start difficult conversations – about work, chores, childcare, religion, responsibility and more. I will never forget the laughter mixed with indignation as women farmers stood up in a group and pointed out that if women were doing so much unpaid labour, then the men should pay them for the work (see McCarthy and Moon 2018). We know that images are powerful for provoking action, for turning citizens into activists (Barberá-Tomás, Castelló, de Bakker and Zietsma 2019). This, combined with the therapeutic dimensions of art-based methods, suggests real promise in VPR for consciousness-raising around a whole range of CSR issues: sustainable transitions, business ethics, citizen activism, social justice and more (McCarthy and Grosser 2020). To what extent this should or could be facilitated by an outsider is, of course, still relevant.

Concluding thoughts

In this chapter, we have reflected upon the promise and pitfalls of using visual participatory methods for empowering research. It is by having the freedom provided by this chapter, as opposed to the constrained structure, conventions and process of a journal article, that we have been able to find our own voices and talk openly about these challenges. Even so, there is always a tension between doing and presenting research. As relatively young female scholars, we often feel the pressure to convince and justify, perhaps more than others may have to.

There is also the unknowable side of what we do. How did our methods affect our participants once we left their office, factory or field? What conversations happened after we prompted discussion on work and home life roles? Might these have been hurtful, or even dangerous? Or perhaps they were the beginning of useful interventions. We will likely never know the answer to these questions, but by reflecting on these issues, we aim to surface some of the successes, challenges and bits-in-between that visual participatory methods provide.

References

Agar, M. H. 1996. *The Professional Stranger: An Informal Introduction to Ethnography*. Bingley, UK: Emerald.

Alcoff, L. 1992. "The Problem of Speaking for Others." *Cultural Critique* 20: 5–32.

Argyris, C., R. Putnam, and D. Smith. 1985. *Action Science*. San Francisco: Jossey-Bass.

Bagnoli, A. 2009. "Beyond the Standard Interview: The Use of Graphic Elicitation and Arts-based Methods." *Qualitative Research* 9(5): 547–570.

Banks, M. 2001. *Visual Methods in Social Research*. London: Sage.

Barberá-Tomás, D., I. Castelló, F. G. de Bakker, and C. Zietsma. 2019. "Energizing through Visuals: How Social Entrepreneurs Use Emotion-symbolic Work for Social Change." *Academy of Management Journal* 62(6): 1789–1817.

Bell, E., A. Bryman, and B. Harley. 2018. *Business Research Methods*. London: Oxford University Press.

Bell, E., N. Kothiyal, and H. Willmott. 2017. "Methodology-as-technique and the Meaning of Rigor in Globalized Management Research." *British Journal of Management* 28(3): 534–550.

Bell, E., S. Warren, and J. E. Schroeder, eds. 2013. *The Routledge Companion to Visual Organization*. London: Routledge.

Bell, P. 2001. "Content analysis of visual images." In *The Handbook of Visual Analysis*, edited by T. Van Leeuwen and C. Jewitt, 12–34. London: Sage Publications.

Bhattacharya, K. 2007. "Consenting to the Consent Form: What are the Fixed and Fluid Understandings between the Researcher and the Researched." *Qualitative Inquiry* 13(8): 1095–1115.

Boyatzis, R. E. 1998. *Transforming Qualitative Information: Thematic Analysis and Code Development*. Thousand Oaks, CA: Sage.

Brannick, T., and D. Coghlan. 2007. "In Defense of Being 'Native': The Case for Insider Academic Research." *Organizational Research Methods* 10(1): 59–74.

Broussine, M. P., and P. Simpson. 2008. *Creative Methods in Organizational Research*. London: Sage.

Budhwar, P. S., D. S. Saini, and J. Bhatnagar. 2005. "Women in Management in the New Economic Environment: The Case of India." *Asia Pacific Business Review* 11(2): 179–193.

Butler, J. 1990. *Gender Trouble*. London and New York: Routledge.

Byrne, A., J. Canavan, and M. Millar. 2009. "Participatory Research and the Voice-centred Relational Method of Data Analysis: Is It Worth It?" *International Journal of Social Research Methodology* 12(1): 67–77.

Caretta, M. N., and Y. Riaño. 2016. "Feminist Participatory Methodologies in Geography: Creating Spaces of Inclusion." *Qualitative Research* 16(3): 258–266.

Chalfen, R. 2011. "Differentiating Practices of Participatory Visual Media Production." In *The SAGE Handbook of Visual Research Methods*, edited by Eric Margolis and Luc Pauwels, 186–200. London: Sage.

Clarke, J. S., and R. Holt. 2019. "Images of entrepreneurship: Using drawing to explore entrepreneurial experience." *Journal of Business Venturing Insights* 11: e00129. https://doi.org/10.1016/j.jbvi.2019.e00129

Davis, C. 2012. "Empowerment." In *The Sage Encyclopaedia of Qualitative Research Methods*, edited by Lisa M. Given, 260–261. London: Sage.

Davison, J., C. McLean, and S. Warren. 2012. "Exploring the visual in organizations and management." *Qualitative Research in Organizations and Management: An International Journal* 7(1): 5–15.

Drew, S., and M. Guillemin. 2014. "From Photographs to Findings: Visual Meaning-making and Interpretive Engagement in the Analysis of Participant-generated Images." *Visual Studies* 29(1): 54–67.

Eldén, S. 2013. "Inviting the Messy: Drawing Methods and 'Children's Voices'." *Childhood* 20(1): 66–81.

Fitzgerald, M., and G. Muske. 2002. "Copreneurs: An Exploration and Comparison to Other Family Businesses." *Family Business Review* 15(1): 1–16.

Fletcher, D. 2010. "'Life-making or Risk Taking'? Co-preneurship and Family Business Start-ups." *International Small Business Journal* 28(5): 452–469.

Gallagher, M. 2008. "'Power Is Not an Evil': Rethinking Power in Participatory Methods." *Children's Geographies* 6(2): 137–150.

Galvaan, R. 2007. "Getting the Picture: The Process of Participation." In *Putting People in the Picture: Visual Methodologies for Social Change*, edited by N. de Lange, C. Mitchell, and J. Stuart, 153–161. Rotterdam: Sense.

Gatenby, B., and M. Humphries. 2000. "Feminist Participatory Action Research: Methodological and Ethical Issues." *Women's Studies International Forum* 23(1): 89–105.

Girei, E. 2017. "Decolonising Management Knowledge: A Reflexive Journey as Practitioner and Researcher in Uganda." *Management Learning* 48(4): 453–470.

Guillemin, M. 2004. "Understanding Illness: Using Drawings as a Research Method." *Qualitative Health Research* 14(2): 272–289.

Harding, S. 1986. *The Science Question in Feminism*. Ithaca: Cornell University Press.

Heinonen, J., and P. Stenholm. 2011. "The Contribution of Women in Family Business." *International Journal of Entrepreneurship and Innovation Management* 13(1): 62–79.

Hinson Shope, J. 2006. "'You Can't Cross a River Without Getting Wet' A Feminist Standpoint on the Dilemmas of Cross-Cultural Research." *Qualitative Inquiry* 12(1): 163–184.

Höllerer, M. A., T. van Leeuwen, D. Jancsary, R. E. Meyer, T. H. Andersen, and E. Vaara. 2019. *Visual and Multimodal Research in Organization and Management Studies*. New York and London: Routledge.

Jackson, C. 2012. "Speech, Gender and Power: Beyond Testimony." *Development and Change* 43(5): 999–1023.

Literat, I. 2013. "A Pencil for Your Thoughts": Participatory Drawing as a Visual Research Method with Children and Youth." *International Journal of Qualitative Methods* 12(1): 84–98.

Maguire, P. 1996. "Considering More Feminist Participatory Research: What's Congruency Got to Do With It?" *Qualitative Inquiry* 2(1): 106–118.

Mama, A. 2007. "Is It Ethical to Study Africa? Preliminary Thoughts on Scholarship and Freedom." *African Studies Review* 50(1): 1–26.

Matzek, A., C. Gudmunson, and S. Danes. 2010. "Spousal Capital as a Resource for Couples Starting a Business." *Family Relations* 59(1): 60–73.

Mayoux, L. 2012. "Gender Mainstreaming in Value Chain Development: Experience with Gender Action Learning System in Uganda." *Enterprise Development & Microfinance* 23(4): 319–337.

McCarthy, L. 2017. "Empowering Women Through Corporate Social Responsibility: A Feminist Foucauldian Critique." *Business Ethics Quarterly* 27(4): 603–631.

McCarthy, L. 2018. "'There Is No Time for Rest': Gendered CSR, Sustainable Development and the Unpaid Care Work Governance Gap." *Business Ethics: A European Review* 27(4): 337–349.

McCarthy, L., and K. Grosser. 2020. "Contesting Hegemonies in Organisational Research Through Consciousness-raising." *Paper presented at the European Group of Organization Studies Conference*, July 6–8th 2020.

McCarthy, L., and J. Moon. 2018. "Disrupting the Gender Institution: Consciousness-raising in the Cocoa Value Chain." *Organization Studies* 39(9): 1153–1177.

McCarthy, L., and J. N. Muthuri. 2018. "Engaging Fringe Stakeholders in Business and Society Research: Applying Visual Participatory Research Methods." *Business & Society* 57(1): 131–173.

McIntosh, P. 2010. *Action Research and Reflective Practice.* London: Routledge.

Meyer, R. E., M. A. Höllerer, D. Jancsary, and T. Van Leeuwen. 2013. "The Visual Dimension in Organizing, Organization, and Organization Research: Core Ideas, Current Developments, and Promising Avenues." *Academy of Management Annals* 7(1): 489–555.

Mohanty, C. T. 1991 "Under Western Eyes: Feminist Scholarship and Colonial Discourses." In *Third World Women and the Politics of Feminism*, edited by C. T. Mohanty, A. Russo, and L. Torres, 51–80. Bloomington and Indianapolis: Indiana University Press.

Oliveira, E. 2016. "Empowering, Invasive or a Little Bit of Both? A Reflection on the Use of Visual and Narrative Methods in Research with Migrant Sex Workers in South Africa." *Visual Studies* 31(3): 260–278.

Opie, A. 1992. "Qualitative Research: Appropriation of the 'Other' and Empowerment." *Feminist Review* 40(1): 52–69.

Pauwels, L. 2015. "'Participatory' Visual Research Revisited: A Critical-constructive Assessment of Epistemological, Methodological and Social Activist Tenets." *Ethnography* 16(1): 95–117.

Pink, S. 2007. *Doing Visual Ethnography.* 2nd ed. London: Sage.

Prosser, J., and A. Loxley. 2007. "Enhancing the Contribution of Visual Methods to Inclusive Education." *Journal of Research in Special Educational Needs* 7(1): 55–68.

Rees, M. 1998. *Drawing on Difference: Art Therapy with People Who Have Learning Difficulties.* London: Routledge.

Reinharz, S. 1992. *Feminist Methods in Social Research.* New York: Oxford University Press.

Ross, K. 2017. "Making Empowering Choices: How Methodology Matters for Empowering Research Participants." *Forum Qualitative Sozialforschung/Forum: Qualitative Social Research* 18(3). http://doi.org/10.17169/fqs-18.3.2791

Shortt, H., and S. Warren. 2019. "Grounded Visual Pattern Analysis: Photographs in Organizational Field Studies." *Organizational Research Methods* 22(2): 539–563.

Silver, R. 2001. *Art as Language: Access to Thoughts and Feelings Through Stimulus Drawings.* London: Routledge.

Spence, L. 2016. "Small Business Social Responsibility: Redrawing Core CSR Theory." *Business and Society* 55(1): 23–55.

Spivak, G. 1988. "Can the Subaltern Speak?" In *Marxism and Interpretation of Culture*, edited by C. Nelson and L. Grossberg, 271–313. Urbana: University of Illinois.

Stiles, D. R. 2013. "Drawing as a Method of Organizational Analysis." In *The Routledge Companion to Visual Organization*, edited by E. Bell, S. Warren, and J. E. Schroeder, 227–242. London: Routledge.

Varga-Dobai, K. 2012. "The Relationship of Researcher and Participant in Qualitative Inquiry: From 'Self and Other' Binaries to the Poststructural Feminist Perspective of Subjectivity." *The Qualitative Report* 17: 1–17.

Vince, R., and M. Broussine. 1996. "Paradox, Defense and Attachment: Accessing and Working with Emotions and Relations Underlying Organizational Change." *Organization Studies* 17(1): 1–21.

Warren, S. 2005. "Photography and Voice in Critical Qualitative Management Research." *Accounting, Auditing & Accountability Journal* 18(6): 861–882.

Zuboff, S. 1988. *In the Age of the Smart Machine.* New York: Basic Books.

8

AUTOETHNOGRAPHY AND PERSONAL EXPERIENCE AS AN EPISTEMIC RESOURCE

Srinath Jagannathan and Premalatha Packirisamy

Introduction

In this chapter, we reflect on the ethical potential of autoethnography as a methodological resource, where textuality is used to explore tensions between the personal and the social (Bhattacharya 2018; Chawla and Atay 2018). Autoethnographic texts as an act and a politics of remembering can help to re-situate identities and break free from ideological constraints (Bell and King 2010; Jagannathan and Packirisamy 2019). Autoethnography recovers the personal as a social performance and helps in overcoming the ideology of responsibilised guilt that can shape the personal. An autoethnographic text is an effort in retrospective sense making and helps in understanding how key experiences are related to the production and contestation of inequality (Haynes 2011; Wall 2006).

As the autoethnographic method draws on personal experience, it becomes necessary to understand the immersion of the personal in conflicting narratives (Bhattacharya 2018). The autoethnographic mode of inquiry can help inquirers to write differently and overcome restrictive norms of scientific writing (Boncori and Smith 2019; Gilmore et al. 2019). In this sense, the recounting of the personal does not follow a linear narrative trajectory. Instead, the recounting of the personal takes a dialectical and conflictual turn as the personal is shaped by numerous conflicts in which it is immersed (Herrmann 2007).

This autoethnographic chapter draws on our ongoing research on the intersections of employment and motherhood, where we engage with Julia Kristeva to understand how abjection shapes the experience of motherhood (Kristeva and Goldhammer 1985; Kristeva 1982). I/Srinath and I/Premalatha fell in love and married while working as colleagues in a university in India. Our relationship has been marked by inequalities of caste and gender. For five years of our marriage, we struggled to have a child. The birth of our child in the sixth year of marriage

DOI: 10.4324/9780429352492-8

generated personal insights into motherhood and fatherhood which we did not imagine previously. In academia, fathers have been shown to experience tensions in reconciling academic work with parenting commitments (Damaske et al. 2014). In order to articulate the production and contestation of inequality, we engage in an autoethnography that connects personal insights arising from the birth of our child and his care to broader political processes. In the autoethnography, we are aware of tensions in our narratives (Jagannathan and Packirisamy 2019). We believe that autoethnography is a methodological resource to enable reflection on how people are marginalised. We reflect on different ways in which marginalisation, personal experience and academic inquiry intersect to produce knowledges which can contest inequality (Guru 2012; Torngren and Ngeh 2018).

Autoethnography can enable hidden stories which are known but not often narrated to become accessible for theorisation (Haynes 2011). At the same time, autoethnographies reveal the precariousness of life (Jagannathan and Packirisamy 2019) and detail its vulnerabilities. Consequently, it takes effort for the autoethnographic narrative to transgress the tropes of a victim narrative (Bhattacharya 2018). While marginalisation and inequality areat play, it must be acknowledged that the narrator is neither a gullible victim nor an actor who can be held responsible for all her experiences (Chawla and Atay 2018). We begin the chapter by outlining our autoethnographic approach. Next, each author presents their narrative of working parenthood independently. Finally, we discuss methodological implications through which personal experiences can be analysed for understanding the production and resistance of inequality, before ending with a brief conclusion.

Crafting autoethnography

Our autoethnography is premised on a lyrical play of memory. Our remembered experiences can be a source of critiquing inequalities in cultures to which we belong (Bell and King 2010). Gilmore et al. (2019, 4) urge inquirers to overcome the limits of scientific writing as scientific tropes excise "much of what it is to be human – the poetics of our humanity . . . and thus our knowledge, understanding and learning are inhibited". In contrast to scientific templates of stability and reliability, our remembered experiences relayered memories with new meanings which were not available earlier. For instance, when we remembered the birth of our child, we were surprised at our courage in going ahead with the birth of our child without seeking support from others.

Lyrical remembering helped us to access narratives that helped us to understand what had happened in our lives. For instance, when we returned from the hospital after the birth of the child, we were overwhelmed with grief at our isolation. We had gone to the hospital as two people and now we returned as three. Nobody else seemed to be bothered about the transition that had occurred in our lives, and we grieved at our isolation. When we remembered the birth of our child, this moment of grief framed many other memories and added layers of meaning not available to us earlier. This is similar to the autoethnographic narrative of Bhattacharya (2018), who suggests that the autoethnographer transfers affect to ordinary events and spaces and acts of remembering are moulded through these affects.

According to Chawla and Atay (2018), autoethnography involves the negotiation of identities in transition. We experienced these liminal markers of identity in our autoethnography as we submitted to the medical authority of the obstetrician. In our autoethnography, we continually shift between multiple identities – as academic, patient, mother, father – and are unsure which identity we are speaking from on many occasions. We looked upon the autoethnographic act of writing as a process of speaking out about experiences that we were uncomfortable with in order to locate our experiences in the political discontents that we wish to articulate.

Our autoethnographic method focuses on the act of writing as the articulation of difference. Our experiences as partners in a marriage with respect to the birth of our child were not the same and involved negotiations of caste and gender. We felt it necessary to articulate these differences in order to acknowledge that our experiences can never be fully shared. While engaging with our differences, we drew from Boncori and Smith (2019) in two ways. First, we were inspired by the ways in which Boncori and Smith give voice to experiences of embodied loss and attempted to articulate our own experiences of grief in a similar vein. Second, we attempted to present fragile, reflexive narratives as they can offer insights for challenging "patriarchal norms of organizing" (Boncori and Smith 2019, 74). We were aware that our fragile narratives were situated in the politics of Indian academia and its ability to censor voices that resist traditional structures. At the same time, we felt that fragile narratives can also become the basis of forming networks of solidarity.

Writing our autoethnographic narrative felt like rearranging different aspects of our grief. The arrangement of our writing was a political performance as events from the past evoked different moments of grief. I/Srinath remembered how intimate relatives had stood in the way of our love and marriage due to reasons of caste and the reproduction of parental authority. My academic employment had been precarious at the time of marriage, and all this shaped my retreat into myself and the loneliness in which my child was born. I/Premalatha became an object of gossip and blame after the marriage. Marriage never proved to be the great escape into a happy life that I had waited for all my life. A mourner's attitude in our writing meant that we were constantly focusing on the absence of social solidarity in the events we were relaying. By situating our writing as an act of mourning, we were engaging with Clifford's (1986) focus on the crisis of representation, and the need to understand the historical and cultural constraints that inform conversations of actors with each other. Both of us mourned differently about our experiences, and our writing was informed by a different set of historical constraints.

Intersecting inequalities and the social exhaustion of motherhood

In my/Premalatha's case, the early death of my parents, a struggle for livelihood and the responsibility of caring for my siblings led to late decisions about love and marriage. I was 43 years of age when I gave birth to my son. The fact that I justify the timing of my motherhood in these terms alludes to the class context of motherhood (Perrier 2013). I was transiting from the working class to the middle class before I was confident of becoming a mother. As we moved towards the birth of our child, Srinath and I were still unsure about our bodies and did not know what the future held for us. Our inexperience in several matters pertaining to the body and sexuality did not help. Co-existence as a negotiation of bodies is a testament of partners in a marriage knowing each other intimately and being ready to care for the other's body.

Finally, to the great joy of Srinath and me, I became pregnant. I wanted to do everything to be safe so that our child was born without any risk. I realise that the discourse of risk is related to the normalisation of medical surveillance of pregnant bodies and women providing reluctant consent to professional control of the process of childbirth (Westfall and Benoit 2008). I was disappointed with the attention that I was getting from obstetricians. Mindlessly, the first obstetrician whom I had visited had ordered tests before the foetus had reached the stage where the test results would be reliable. On expected lines, I received unreliable test results indicating high risk pregnancy. Instead of acknowledging her error and advising a retest when the foetus would have evolved, she recommended another intrusive surgical test which itself had a small proportion of risk attached to it. This obstetrician had won several awards and was a famous figure in the country. She was regarded as a pioneer in In Vitro Fertilization (IVF) assisted births. I felt that she did not pay that much attention because mine was a spontaneous conception and not a case of IVF assisted birth. IVF assisted births are far more financially lucrative to doctors than spontaneous conceptions. During one of the examinations, the obstetrician crossed her fingers and remarked to her junior colleagues, "elderly prime". I felt she was acting with prejudice about my age and this made me feel despondent. Submitting myself to obstetrical authority when the gynaecologists did not particularly seem to care for me as a person was disturbing.

Medical examination and assistance have taken a largely commercial turn in India. In the context of this commercial turn, it becomes difficult to trust the obstetrician. One of the things that helped me cope with the alienation of medical authority was joining a yoga group for pregnant women. It provided me with a peer group with whom I could discuss a range of issues. It also helped me to counteract the medical discourse of risk and sustained the belief that everything was alright with my body. It built a sense of confidence and hope that I would be able to have a normal delivery and would not need a c section.

Within the framework of liberal capitalist healthcare, the assertive consumer is expected to shop around for the best healthcare possible by exercising her choice (Collyer et al. 2015). Yet as Collyer et al. indicate, the vulnerability of patients prevents them from behaving as assertive consumers. Both Srinath and I were seen as soft-spoken people who would not be able to stand up for their rights as consumers. I felt that our niceness was exploited. Even on the day when I was giving birth to my son, the obstetrician came late by almost an hour. She was overseeing another C- section delivery for almost an hour after I was ready to give birth to the child at any moment. The team of junior doctors, midwives and nurses were requesting me not to push as the child could be born at any moment. The midwife commented to the junior doctor, "Where is the obstetrician? She should come immediately. The head of the child is visible". I felt that the other patients were more aggressive and the obstetrician decided to attend to them rather than come to begin my delivery forthwith. This was also borne out during my earlier medical check-ups with the obstetrician. Just as my turn for the examination would come up, the obstetrician would make an excuse and go away on another errand. On many occasions, in comparison to other patients, I had to wait longer for the examination.

Due to the delay, while a normal delivery could still occur and a C-section was not needed, my son had passed stools while being born. The delay indicated that private sector health care based on user payments does not necessarily lead to better outcomes (Collyer et al. 2015). I believe that the little one panicked inside the womb due to all the contractions and activity around him. After the delivery, I was hoping that the staff would help me to take a bath and clean myself. One of the staff members told me that everything was over and there was no need to take a bath. I felt exhausted after the process of birth and a bath would have satisfied me. I was disappointed that neither the obstetrician nor other staff were concerned about my voice.

Due to the infection of being stained by his stools, my son had developed symptoms of laboured breathing. After a couple of hours, they took my son away to the neonatal intensive care unit (NICU). At that time, I did not know what was happening. I thought they will bring back my son in a couple of hours. But now when I reflect on what happened, I feel agitated about how commerce transforms a mother's body into a site of business. When the obstetrician came late and my son developed an infection, it turned out to be a more profitable venture for the hospital as the NICU charges were more than the usual charges. I realise that structural constraints operate in hospitals which make doctors and patients behave in ways signifying inequality (Collyer et al. 2015).

One of my friends who had her delivery in the same hospital had warned me that the quality of post-birth paediatric care was poor and could be premised on commerce. While her daughter had not developed jaundice, the hospital reported to her that the newborn child was afflicted with jaundice. This was an attempt to inflate the hospital bill. It was only when my friend resisted and read through the medical reports carefully that the hospital relented, apologised and rolled back the bill. While I had this in mind and was not sure whether my son needed to spend four days in the NICU, I could not resist the medicalisation discourse. My apprehensions about commerce were heightened when the obstetrician asked me whether I had medical insurance. While admitting myself as a patient, I had consciously decided not to report insurance. Srinath had declined to use the medical insurance available from his employer. My employer's medical insurance would ensure the reimbursement of only a small fraction of the costs. I was disappointed that doctors would resort to different practices of care depending on the insurance available with patients. In India, there is a tendency among hospitals to inflate the expenses when medical insurance is available.

The discourse of medicalisation which leads to women feeling a loss of autonomy over their bodies is premised on a collapse of collegiality among the staff delivering healthcare (Westfall and Benoit 2008). When I went to the NICU to meet my son, one of the nurses taking care of the child asked me whether I had a spontaneous conception. She told me that she had been trying for a child with her spouse for about five years and was yet to conceive. She indicated that they were used to receiving children of lesser weight in the NICU and my son had the weight of a full-term baby. They were using diapers of a smaller size compatible with children of lesser weight. The nurse admitted that the smaller diapers were creating problems, but yet did nothing to order bigger diapers that would fit my son's size.

When they took away my child to the NICU, I found it difficult to breastfeed my baby in the NICU with all sorts of wires being attached to him. I also had to wear protective gowns and masks while entering the NICU which added to the difficulties. The nurse in the NICU squeezed my breasts roughly to try and help me in breastfeeding my son. What the nurse did was very painful, and I felt that it was not helping me in taking care of my son. I experienced a sense of being bullied by the nurse. I wondered whether this was related to her own anxiety of not having a child. As a result of my not being able to breastfeed him, my son got used to drinking milk from the feeding bottle. In the meantime, the hospital sent another nurse who was a lactation consultant to visit me. She spent a few minutes with me and I learnt nothing useful from her. But I noticed that her visit was added to the hospital bill.

I had no experience of how a mother's body behaved and I found that the hospital staff were not forthcoming. I was developing some pain in my breast a day after my child's birth. No one seemed to notice. Accidentally, a nurse came to know about my pain and told me that milk was accumulating in my breast and I should buy a pump and immediately express milk. This expressed milk could be given to my son. Nobody had alerted me before this that milk was likely to accumulate in my breast and I should be prepared to express this. In the initial days, either I used to express my milk and give it to my son using the bottle or milk was prepared using a commercially sold formula. Due to the initial habits he had developed, it took my son almost a month to get used to breastfeeding. During this period, Srinath felt that in order to provide the child with nutrients, it was necessary to rely more on formula milk. I felt that unless the reliance on formula milk was reduced, the child would not easily take to breastfeeding.

During this process, I was again reminded of the inadequacy of my body. My colleagues and others suggested that due to my age I was not generating enough milk. Earlier, almost right till I went into labour, a senior woman colleague kept suggesting that I would not be able to have a normal delivery and would have to go for a C-section. I felt she was almost casting a curse on me. During the same time, I had become chairperson for my centre and colleagues thwarted several interventions I was proposing. They resisted my responsibilities in various ways, and overall, I found hostility in the workplace when I was pregnant. Grossman (2016) contends that these experiences of hostility emerge from the unwillingness to treat the pregnant body as a body that is in waiting, a body that is being severely taxed by the life it is nourishing. In the process, I complied with regular labour in terms of teaching and other responsibilities. I did not take any leave before the birth of the child. I commenced my maternity leave only after my son was born.

During my stay in the hospital, one of the cleaners frequently asked me why nobody from our homes had come to help us. I had been advised to take a sitz bath while being in the hospital, and though it was part of her job, the cleaner was reluctant to help me. The cleaner described the difficulties she had gone through while giving birth to her child and how she had returned to work immediately afterwards. I felt that the cleaner was envious that I was having a more comfortable existence after my childbirth and not experiencing the same degree of pain she had experienced. This increased my disappointment about the care I received in the hospital. On one occasion, Srinath had to go back home in the morning to bring a new set of clothes. At that time, the cleaner and the nurses asked me whether Srinath had left me alone in the room and whether there was no one else from either of our families around. I felt that the hospital staff were not willing to take the responsibility of caring for me and preferred others from the family to care of me. All this did not prevent Srinath and me from generously tipping the cleaners in the labour room and the recovery room, and buying chocolates for nurses, doctors and my yoga teacher.

Though I was unhappy with the hospital during the birth of my child, I was privileged and fortunate to receive the care I did. There are many in India who do not have access to basic reproductive care because they cannot afford to access health care. The public health delivery system in India produces inequalities where reproductive care is not available to those who need it most. This leads to problems of high maternal mortality and under-weight babies in India (Coffey and Hathi 2016).

I felt that the gendered discourse of the maternal body being an irrational and uncontrollable entity is internalised by medical expertise. The subordination of the woman's body is in line with the project of modernity which intends to subordinate the sensuous to the will of the intellect and reason (Kristeva 1982). Doctors view the maternal body as a diseased body that needs to be regulated. If doctors are nice to the patient, it is due to the force of capital, which urges the doctor to be nice to the consumer-patient. When the first obstetrician whom I visited suggested an invasive investigation, and the second obstetrician insisted on a non–invasive test, they were looking upon my body as an anomaly. It is necessary to restore the discursive primacy of craft in the enactment of medicine to enable doctors to think about the maternal body as something other than a diseased body.

In the case of my child, Srinath and I engaged with the birth of the child alone. There was no one else around to welcome the child. The absence of the act of welcoming meant that after giving birth to the child, I was exhausted. I heard my child's loud grunting sounds after he was born. I noted the fact that he had been taken away to the NICU, and I could not even look at him properly before they took him away to the NICU. In my exhaustion, I could only eat and sleep. As our boy was taken away to the NICU, Srinath and I had a good meal. When we returned to our room from the labour room, we ordered some more food and ate it. Our exhaustion indicated how the lack of solidarity during childbirth can lead to a sense of being tired. As an informant noted in Westfall and Benoit's (2008, 66) study, when people draw together during the time of pregnancy, the act of solidarity leads to joy and reduces the exhaustion of the mother.

When I went into the NICU, I felt harrowed to see all sorts of wires covering my son. But a sense of love and grief had not yet taken over me. It would take at least a couple of days for the sense of love to overwhelm me. For the moment, I was just exhausted and fell asleep. The exhaustion was related to the medicalisation of my pregnancy and the anxiety I was made to go through from the beginning. My sense of exhaustion was related to a range of factors. From the beginning, my engagement with obstetricians had been alienating and I was looking for a closure to my consultations with them. Colleagues in my workplace had been fairly hostile during the course of my pregnancy. My marriage with Srinath, with all its ups and downs, had isolated us from a larger sense of family. I had been anxious that my child should be born safely and through a normal delivery so that I could recover fast and take care of my child soon. I realise that I had allowed the medicalisation of my pregnancy to take me over in many ways. As Westfall and Benoit (2008) indicate, the medicalisation discourse inserts pregnancy into a risky narrative and it is only in retrospect that people realise that everything was normal. It was ironical that I got more information about pregnancy from Srinath's and my friends than from doctors.

Studies have looked at motherhood from various prisms such as age, choice of care during the birth of the child and inequality in accessing health care for pregnant women and mothers (Coffey and Hathi 2016; Grossman 2016; Perrier 2013; Westfall and Benoit 2008). Drawing from Kristeva (1982), the maternal body has been described as occupying an abject imagination which is simultaneously fascinating and repulsive. I add to these theoretical strands by arguing that the abjectness of motherhood emerges from a range of inequalities, including the erosion of collegiality in the workplace. The medicalisation of pregnancy acts in intersection with social inequalities to structure a sense of exhaustion for the mother. It becomes difficult to immediately recognise motherhood as an act of love, as the process of childbirth is exhausting.

My main contribution in the reflection through this autoethnography is that women's routes of pregnancy and motherhood are complicated. Even within a heterosexual marriage premised on the imagination of love, women have to assert themselves and be quarrelsome in order to make their spouses behave in civil ways. Motherhood can become an individualised process completely devoid of social expressions of solidarity when a range of inequalities come together. I contend that it is necessary to reconstitute motherhood as a social act with solidarity being necessary to repair the damages of the medicalisation of the body. In the absence of motherhood becoming a social process, the experience of pregnancy, childbirth and early care of the child leaves the mother not merely physically but socially and culturally exhausted.

Recovering fatherhood

I/Srinath found that caste operates as the anti-thesis of love. Caste is premised on the logic of segregation and the reproduction of inequality (Gorringe and Rafanell 2007; Jagannathan, Bawa and Rai 2020; Mahalingam, Jagannathan, and Selvaraj 2019). People in my intimate kinship network opposed my pursuit of love and used all possible strategies to prevent me from marrying Premalatha. They resorted to emotional blackmail stating that they would be adversely affected if I went ahead with the decision to marry Premalatha. They raised formal objections in the registrar's office opposing our marriage. They engaged in defamatory tactics by calling several colleagues to express their resentment about our marriage. They bullied and humiliated Premalatha to dissuade her from marrying me. The humiliation that my relatives heaped on Premalatha is in line with how Dalits have to resist discrimination and inequality on an everyday basis to reclaim dignity (Gundimeda 2016; Jagannathan, Packirisamy and Joseph 2018; Jagannathan, Selvaraj and Joseph 2016). The lens of caste drives people to become antithetical to love. In order to defend caste, people believe that it is necessary to censor love. The censorship of love has a lot of social support in India. Many of our colleagues frowned on our marriage and felt sympathy for people in my intimate kinship network.

My initial days of marriage were marked by the precariousness of my employment relationship. Precarious employment relationships are a part of academia all over the world and mirror the insecurities of project-based work that inform the employment relationship in several contexts (Zawadzki and Jensen 2020). For people with PhDs in India as in other parts of the world, employers who offer spaces where academic workers can engage in contemplation, dialogue and intervene in society in the interests of justice and equity are increasingly rare. I was not sure whether my contract would be renewed in the university. A lot depended on how people thought I behaved. For my contract to be renewed, I had to establish myself as a sufficiently docile figure who would not disturb the turfs of the administration or established academics. Increasingly, academia is characterised by hierarchical managerialism within universities which structures academia as subordinate to industrial capitalism (Parker 2014). A powerful academic in the university had taken it upon himself to ensure my contract was not renewed. In order to find reasonable employment, I would have to move to another city as not many academic jobs were available in the city in which the university was located. Even when I got a job in another city in another institute, all that I could manage was another temporary position.

Precarious employment contracts are about the temporality of dislocation. Within universities, flexibilised work arrangements have been increasing, leading to experiences of precariousness, overwork and dissatisfaction (Allmer 2018). My precarious employment contract meant that I struggled to hold on to my job. Wilson and Ebert (2013) have argued that precarious jobs cause social and economic distress and affect the quality of family life. Along with my struggle in a precarious job, Premalatha and I struggled as lovers and married partners. We did not have a child for a long period of time into our marriage and there were resulting tensions. Identifying myself with the politics of critical management studies also made me ambiguous about having a child. With so much inequality around the world, I did not know what a child would do by coming into the world. I was opposed to the idea of inheritance. Even when I was on good terms with my parents, I had told them I would not be inheriting the wealth they left behind. With a child coming into the world, many such questions of privilege and inequality would follow. Fatherhood is a political project connected with socialising the child into the space of citizenship. Children often continue with the same political allegiances that their fathers have (Veugelers 2013).

Ba' (2019) contends that precarious employment relationships characterised by everyday struggles for stability introduce vulnerability into the process of parenting. Parental vulnerability is related to a broader crisis of care where it is increasingly becoming difficult to mobilise resources for the care of children and older people (Alberti et al. 2018). However, by the time my child arrived, I had moved away from the space of precarious employment to a more stable academic job. Today, I am able to contest the unilateralism of administrative actions, resist indignities, claim flexible working hours and spend more time with my son. In a precarious employment relationship, I may have been more careful in engaging with the administration, may have tolerated indignities and may not have been able to claim flexible working hours. I may have had to maintain more regular office hours and give up on caring for my son. As a precarious academic worker, I may have been made to feel that I did not deserve to have a son and begin a family. Precariousness can lead to gendered consequences such as fathers being unable to contribute to the care of their children, resulting in unequal sharing of parental labour.

While a powerful figure in the university was responsible for my precariousness in important ways, he was a progressive sociologist in the country. True to his commitment to a progressive cause, he welcomed our marriage and congratulated us warmly. That a progressive sociologist can be responsible for structuring precariousness for junior academics is a testament to the nature of academic spaces. Zawadzki and Jensen (2020) contend that the neoliberal academic context has enabled the bullying of junior academics who face precariousness and intense workloads in universities. Zawadzki and Jensen (2020, 2) contend that "opportunism, authoritarian dependency . . . a lack of autonomy, pauperization of junior scholars, the ghettoization of their research and the marketization of the idea of meritocracy" characterise the university in current times.

Academic spaces can benefit from deterritorialisation and resulting plurality. According to Deleuze and Guattari (1987), territorialisation of space is related to inserting order and regulation into space in contrast to allowing space to take turns of multiplicity and distributedness. With the increasing marketisation of academic spaces, powerful academics use processes of the market to cement the precariousness of other academic actors. They demand allegiance from academic actors to help them navigate the market and build a layer of security for themselves. The identities of academics are constructed by the social relations of production in the academia which influence how academics engage with each other and the nature of academic outcomes that are produced (Zawadzki and Jensen 2020).

I could not get a decent academic job in the city in which the university was located and had to relocate to another city. I have a significant number of classes and work in another city and am not able to be with my son always. I also had to take up the chairpersonship of a new programme for my institute, which involved a lot of 'selling' the programme to corporations to generate internship and job opportunities for students. This is in line with neoliberal agendas that are shaping contemporary universities where universities compete with each other to attract students in order to remain sustainable (Zawadzki and Jensen 2020). In order to attract students, it becomes necessary to promise students that they will be able to obtain the best jobs available in the market after graduating. This act of 'selling' subordinated me as an academic to corporate actors as it established the subservience of academia to the market. Also, in building these connections, I engaged in self-exploitation as I spent a lot of personal money to nurture the programme.

Engaging in marketing an academic programme is tiring and corrosive. Increasingly, academia has taken a marketing turn with academic knowledge being commodified and the logic of accumulation and growth becoming central to academic institutions (Lincoln 2012). I felt that I was beginning to look at everything from the lens of an opportunity that could be leveraged for the academic program. Further, the act of marketing stood in contrast to the commitment to critical management studies that I had. These contradictions consumed me and left me tired in engaging with my child. Precariousness and market-based transformation of academia took away my joy of being an academic worker.

Fatherhood appeared to me in the form of scheming and manoeuvring as I sought to be with my son as far as possible. This is in line with what Ekman (2013) describes as the high involvement worker who invests a lot of herself and time in her work being both precarious and opportunistic. The need to reach out to corporate stakeholders meant that my care for my son was disrupted. After teaching classes in another city during the week, I had to go on Sundays to the campus in the city in which my family lived to facilitate guest talks by corporate speakers. The time I could spend caring for my child was being taken away from me. Fatherhood requires a reorientation of academic work, but there is little support available for such reorientation.

I was surviving the job I was doing and returning home for refuge from my work. In many ways, my son was caring for me than the other way around. I had initially joined academia imagining a romantic, poetic, intellectual, dialogical community. But the romantic imaginary of the university, related to democratic knowledge and identities, is increasingly on the wane due to market forces (Zawadzki and Jensen 2020). Even when I found relative stability in an academic job, the marketing orientation that my job brought destroyed any imagination of a romantic community. The only romantic community was available at home, and I had to protect my time from being eaten away by my academic job. Paradoxically, I was using metaphors of sovereignty and security when I talked about protecting my time and locating a romantic community at home with Premalatha and my son.

Previous work on precariousness and parenting has outlined how precarious employment relationships create a crisis of care and structure struggles for parents (Alberti et al. 2018; Ba' 2019). This crisis of care is compounded by working in academia where market forces are shaping academic work in commodified forms, taking the romantic and poetic edge away from academia (Allmer 2018; Lincoln 2012). Working within structures of commodification ethically consumes academic workers altering the orientation of care towards their children. In a gendered sense, work-related pressures in academia could become an excuse for fathers to reduce their share of labour in caring for children (Damaske et al. 2014). The crisis of care is accentuated by social relations of caste which break down the possibility of solidarity and normalise inequality (Gorringe and Rafanell 2007; Gundimeda 2016).

Emerging from this discussion, the theoretical focus of the intersection of precarious work, academic labour, fatherhood and caste has been on how a range of factors come together to prevent care towards the child being enacted in effective ways. I add to this literature by arguing that caste, precarious work and academic labour not only come in the way of enacting fatherhood as a labour of care but also come in the way of practicing fatherhood as a labour of love. Love needs to be rescued from the inequalities of caste and academic labour. Caste and the market transformation of academic labour can deeply scar subjects and inhibit their capacity to love. The ability to immerse fatherhood in the poetic practice of love can emerge only by romantically resisting caste and market forms of academic labour in as many ways as possible.

Autoethnographic contributions

There are primarily two contributions that we make through our autoethnographic inquiry. First, while various studies have indicated how inequality informs the experience of motherhood (Coffey and Hathi 2016; Grossman 2016; Perrier 2013; Westfall and Benoit 2008), we focus on how the intersection of life experiences and the medicalisation of the pregnant body transform motherhood and the birth of the child into a precarious experience. I/Premalatha experienced the medicalisation process in alienating ways as the delay in my son's delivery led to him being infected and spending time in the NICU. Hostility continued in implicit ways as colleagues, friends, Srinath and domestic workers suggested that I would not be able to adequately breastfeed my child due to my age.

Second, while earlier studies have pointed to a range of factors such as inequality and precarious work which come in the way of fathers providing care to the child (Alberti et al. 2018; Ba' 2019), we argue that the intersection of caste and the market transformation of academic labour come in the way of fatherhood being seen as an act of love. Rather than an act of love, fatherhood becomes an act of protecting the time between the father and the child from being stolen away by other entities. The experience of having to protect the time with one's child becomes an act of securitisation rather than one of love. Securitisation is about a sense of fear that the markers of caste will socialise the child in scarred ways, and the need to protect the child from the extended family and other social actors who are seen as purveyors of caste. While I had not informed my parents about the birth of my child, I/Srinath have told them about my son at the time of writing this chapter. While my parents have taken to my son with some affection, they do so only as my son is not yet a speaking, politically rebellious being. At the time of writing this chapter, my father passed away due to a heart attack in late April 2020.

I/Srinath only had an abstract idea about how caste censored practices of love (Jagannathan, Bawa and Rai 2020; Jagannathan and Packirisamy 2019). But when I rebelled against my parents to marry Premalatha, I personally understood how caste is the antithesis of love. Similarly, I had only an abstract idea of how neoliberal discourse structured an enterprising self that was drawn into self-exploitation and precariousness. I had read about how academics experienced a change in the context of academic labour and how the market had transformed academia into a place where romantic imagination was on the retreat (Parker 2014). Through my personal experience, I acquired meaningful knowledge about how discourses of marketing and enterprise created corrosive impacts. Fatherhood in the midst of a neoliberal practice of academic labour could only be crafted as a negotiation of escape.

Premalatha and I had no support in taking care of our child. We were reluctant to entrust any part of the care to someone else in lieu of wages as we felt that they would claim an expert authority, which would be difficult for us to challenge. According to Christopher (2012), this falls within the paradigm of intensive mothering where the mother wants to take up as much responsibility for the care of the child as possible and is reluctant to engage paid labour for the care of the child. Whenever my academic labour took me away, Premalatha was alone caring for the child. I did not want to burden her, and so it became a game of escaping from the demands of my academic labour to spend as much time at home as possible. Also, fatherhood was a journey for me to learn the skill of caring for my child.

Initially, when we brought the child from the hospital, I was hopelessly inadequate. I did not know how to hold the child in my hands. Over time, I learnt to engage with my son's body. I learnt film songs to sing to him. I was beginning to understand how the social and occupational world in which I was situated was not open to supporting a father's love for his child. As Damaske et al. (2014) indicate, male academics find that demands of academic careers are in conflict with their acting in egalitarian ways at home in caring for the child. My act of love was located in a struggle against many social and occupational forces, and I was not always successful in this struggle.

Locating knowledges in personal experiences

Both of us realised that we had obtained insights about motherhood and fatherhood as a result of our own experiences that we would not have been able to obtain otherwise. According to Guru (2012, 127), it is useful to acknowledge that knowledges are not absolute or stable and thereby consider experience as an epistemic resource for theorisation. Writing premised on experience can draw from poetry, dreams and imagination (Gilmore et al. 2019). Knowledges based on personal experience can produce uncomfortable moments for the reader as the text proceeds to a promised grief and the reader is engaging in an act of solidarity with the narrator to empathise with the grief (Boncori and Smith 2019). Personal experience adds additional fragmentations to imaginations that people have about phenomena. The autoethnographic gesture of personal experience draws courage from a feminist turn of knowledge, a tradition of memoir, which has "a long record of fearlessly subverting academic discourse, including by inserting embodied selves" (Sinclair 2019, 145).

We believe that knowledges take an ethical turn when they enable breaking free from concerns of efficiency, optimality and enterprise to question how social and organisational constraints placed on actors marginalise them. According to Guru (2012, 128), the ethical turn of theory is premised on respecting experience and contesting the binary of experience formulated as emotion and theory conceived as reason. Often, efficiency, optimality and enterprise are fictions which are carefully nurtured by hegemonic actors to protect their interests. In the case of academia, concerns of enterprise are a discourse that enables the subordination of academia to the imperatives of industrial capital. The concerns of enterprise take away from academics the possibility of fulfilling experiences, such as motherhood and fatherhood, immersed in social solidarity and love.

Marginalisation and responsibilisation

The inquirer's personal immersion in the phenomena can help in the ethical turn of knowledge by engaging with the production of marginalisation. Interpreting the production of marginalisation is not straightforward even in autoethnographies. Bell and King (2010) pay attention to the process of marginalisation by outlining how masculinities become normalised in critical spaces. Engagement with the production of marginalisation involves the reversal of responsibilisation. Responsibilisation implies that actors begin to believe that their choices and actions are responsible for outcomes and other structural factors matter little for these outcomes (Foucault 2008). For instance, the medical discourse largely responsibilises the pregnant woman for any childbirth-related outcome.

Similarly, with the language of enterprise informing academia, the academic is made to feel responsible for meeting market-based targets (Allmer 2018). In our autoethnography, we contest both these projects of responsibilisation and outline that the medical project of expertise which responsibilises the pregnant woman during childbirth contributes to social exhaustion. Similarly, the project of an academic needing to be enterprising marginalises the possibility of love in the space of fatherhood. When the inquirer's personal immersion contests discourses of responsibilisation, the inquirer can enable the contestation of hegemonic consent. Through an autoethnographic gesture, the inquirer may be able to disidentify with dominant structures and tropes and engage in the discovery of a new mode of voice (Chawla and Atay 2018).

We believe autoethnographies can provide important clues about the social relations of love that can help in contesting unequal social structures. Love involves the discovery of vulnerability, pain, angst and the acknowledgement of the lyricality of experiences with respect to others, a move from the individual towards the common. The discovery of the willingness and joy of collectively navigating vulnerability as an experience of love can threaten prevailing social orders which censor the solidarity emerging from love. According to Kiriakos and Tienari (2018, 263), love constitutes

> action rather than feeling . . . reflects multifaceted experiences . . . confusing and disrupting the masculine order . . . offers a language to talk about vulnerability and courage . . . helps us to learn more about ourselves and our activities.

Love signals a move away from the plane of interpersonal feeling to action immersed in the commons.

My/Premalatha's love towards my child signified several aspects of commons beyond the medicalisation of motherhood. The medicalisation of motherhood constitutes the mother and the child as medical objects governed by scientific interpretations of the body. These scientific interpretations of the body intersect with practices of commerce and authority to mark the experience of motherhood with discourses of risk. In searching for craft and solidarity beyond the medical discourse, I was looking for a commons of care that would help the mother and the child navigate the world. The commons of solidarity would replace hostilities of various sorts such as the ones which suggested that my body was inadequate in breastfeeding my child with expressions of care.

My/Srinath's actions of love towards my child signified several aspects of commons rather than a narrow, interpersonal domain of affection. I could express better care towards my child if academia embodied a spirit of commons and solidarity than a sense of enterprise. If I bought into the ideology of enterprise and did not resist it, I would become a responsibilised actor within academia and would find it difficult to enact love towards my child. Similarly, I wanted love towards my child to be enacted in the commons of equality and the resistance against caste. As a result, dilemmas arose for me about how I would enable my child to engage with people in my kinship network, including my parents, in the interpersonal domain. I recognised that my parents' interpersonal love for their grandson could stand in tension against a commons of equality and the resistance against caste. In my actions, I had to navigate these dilemmas to embed love in the sensibility of actions than in the realm of interpersonal feelings.

Conclusion

Positivist scientific knowledge has marginalised personal experience as an epistemic resource by labelling personal experience as being unstable and volatile. The auto-turn in knowledges has confronted the ideology of positivist science by grounding knowing in personal experience, giving voice to the margins and breaking silences. Autoethnography has the potential to ethicalise knowledges by outlining additional fragmentations in local contexts which could not have been imagined earlier. The additional layers of fragmentation pluralise the phenomena and outline how a range of factors and fictions are at play in the actuation of the phenomena. The ethical turn in autoethnography helps in understanding a greater breadth of issues that operate in local contexts.

We need to be aware that the autoethnography is a stitching together of fictions, but these fictions take an ethical turn when they begin to confront ideologies. For instance, it is a fiction that I/Srinath will be able to love my child if the constraints of neoliberal academic enterprise and the inequality of caste did not exist. But the fiction of love is confronting the ideology of caste and academic enterprise. Similarly, I/Premalatha know that it is a fiction that by accessing expressions of solidarity during childbirth, the maternal body will be less socially exhausted. But this fiction of solidarity contests acts of individualisation which make the mother feel isolated in caring for the child. In this chapter, we hope that we have been able to outline an autoethnographic move that can resist ideologies which seek to responsibilise actors and make them politically inert.

References

Alberti, G., Bessa, I., Hardy, K., Trapmann, V. and Umney, C. (2018) In, Against and Beyond Precarity: Work in Insecure Times. *Work, Employment and Society* 32(3): 447–457.

Allmer, T. (2018) Precarious, Always-on and Flexible: A Case Study of Academics as Information Workers. *European Journal of Communication* 33(4): 381–395.

Ba', S. (2019) The Struggle to Reconcile Precarious Work and Parenthood: The Case of Italian 'Precarious Parents'. *Work, Employment, Society* 33(5): 812–828.

Bell, E. and King, D. (2010) The Elephant in the Room: Critical Management Studies Conferences as a Site of Body Pedagogics. *Management Learning* 41(4): 429–442.

Bhattacharya, K. (2018) Coloring Memories and Imaginations of 'Home': Crafting a De/colonizing Autoethnography. *Cultural Studies ←→ Critical Methodologies* 18(1): 9–15.

Boncori, I. and Smith, C. (2019) I Lost My Baby Today: Embodied Writing and Learning in Organizations. *Management Learning* 50(1): 74–86.

Chawla, D. and Atay, A. (2018) Decolonizing Autoethnography. *Cultural Studies ←→ Critical Methodologies* 18(1): 3–8.

Christopher, K. (2012) Extensive Mothering: Employed Mothers' Constructions of the Good Mother. *Gender and Society* 26(1): 73–96.

Clifford, J. (1986) Introduction: Partial Truths. In *Writing Culture: The Poetics and Politics of Ethnography*, edited by James Clifford and George E. Marcus, 1–26. Berkeley: University of California Press.

Coffey, D. and Hathi, P. (2016) Underweight and Pregnant: Designing Universal Maternity Entitlements to Improve Health. *Indian Journal of Human Development* 10(2): 176–190.

Collyer, F.M., Willis, K.F., Franklin, M., Harley, K. and Short, S.D. (2015) Healthcare Choice: Bourdieu's Capital, Habitus and Field. *Current Sociology* 63(5): 685–699.

Damaske, S., Ecklund, E.H., Lincoln, A.E. and White, V.J. (2014) Male Scientists' Competing Devotions to Work and Family: Changing Norms in a Male Dominated Profession. *Work and Occupations* 41(4): 477–507.

Deleuze, G. and Guattari, F. (1987) *A Thousand Plateaus*. Minneapolis: University of Minnesota Press.

Ekman, S. (2013) Is the High-Involvement Worker Precarious or Opportunistic? Hierarchical Ambiguities in Late Capitalism. *Organization* 21(2): 141–158.

Foucault, M. (2008) *The Birth of Biopolitics: Lectures at the College de France, 1977–1978*. New York: Palgrave Macmillan.

Gilmore, S., Harding, N., Helin, J. and Pullen, A. (2019) Writing Differently. *Management Learning* 50(1): 3–10.

Gorringe, H. and Rafanell, I. (2007) The Embodiment of Caste: Oppression, Protest and Change. *Sociology* 41(1): 97–114.

Grossman, J.L. (2016) Hard Labor: The Pregnant Body at Work. *Law, Culture and the Humanities* 12(3): 466–473.

Gundimeda, S. (2016) *Dalit Politics in Contemporary India*. Abingdon: Routledge.

Guru, G. (2012) Experience and the Ethics of Theory. In *The Cracked Mirror: An Indian Debate on Experience and Theory*, edited by Gopal Guru and Sundar Sarukkai, 107–127. New Delhi: Oxford University Press.

Haynes, K. (2011) Tensions in (Re)presenting the Self in Reflexive Autoethnographical Research. *Qualitative Research in Organizations and Management* 6(2): 134–149.

Herrmann, A.F. (2007) How Did We Get This Far Apart? Disengagement, Relational Dialectics and Narrative Control. *Qualitative Inquiry* 13(7): 989–1007.

Jagannathan, S., Bawa, A. and Rai, R. (2020) Narrative Worlds of Frugal Consumers: Unmasking Romanticized Spirituality to Reveal Responsibilization and De-politicization. *Journal of Business Ethics* 161(1): 149–168.

Jagannathan, S. and Packirisamy, P. (2019) Love in the Midst of Precariousness: Lamenting the Trappings of Labour in De-intellectualized Worlds. *Decision* 46(2): 139–150.

Jagannathan, S., Packirisamy, P. and Joseph, J. (2018) Worlds of Demonetisation and Delegitimising the Grief of the Marginal. *Journal of Marketing Management* 34(11–12): 965–988.

Jagannathan, S., Selvaraj, P. and Joseph, J. (2016) The Funeralesque as the Experience of Workers at the Margins of International Business: Seven Indian Narratives. *Critical Perspectives on International Business* 12(3): 282–305.

Kiriakos, C.M. and Tienari, J. (2018) Academic Writing as Love. *Management Learning* 49(3): 263–277.

Kristeva, J. (1982) *Powers of Horror: An Essay on Abjection*. Translated by Leon S. Roudiez. New York: Columbia University Press.

Kristeva, J. and Goldhammer, A. (1985) Stabat Mater. *Poetics Today* 6(1/2): 133–152.

Lincoln, Y.S. (2012) The Political Economy of Publication: Marketing, Commodification, and Qualitative Scholarly Work. *Qualitative Health Research* 22(11): 1451–1459.

Mahalingam, R., Jagannathan, S. and Selvaraj, P. (2019) Decasticization, Dignity and Dirty Work at the Intersections of Caste, Memory and Disaster. *Business Ethics Quarterly* 29(2): 213–239.

Parker, M. (2014) University, Ltd: Changing a Business School. *Organization* 21(2): 281–292.

Perrier, M. (2013) No Right Time: The Significance of Reproductive Timing for Younger and Older Mothers' Moralities. *Sociological Review* 61(1): 69–87.

Sinclair, A. (2019) Five Movements in an Embodied Feminism: A Memoir. *Human Relations* 72(1): 144–158.

Torngren, S.O. and Ngeh, J. (2018) Reversing the Gaze: Methodological Reflections from the Perspective of Racial and Ethnic Minority Researchers. *Qualitative Research* 18(1): 3–18.

Veugelers, J.W.P. (2013) Neo-Fascist or Revolutionary Leftist: Family Politics and Social Movement Choice in Postwar Italy. *International Sociology* 28(4): 429–447.

Wall, S. (2006) An Autoethnography on Learning about Autoethnography. *International Journal of Qualitative Methods* 5(2): 1–12.

Westfall, R. and Benoit, C. (2008) Interpreting Compliance and Resistance to Medical Dominance in Women's Accounts of Their Pregnancies. *Sociological Research Online* 13(3): 62–77.

Wilson, S. and Ebert, N. (2013) Precarious Work: Economic, Sociological and Political Perspectives. *Economic and Labor Relations Review* 24(3): 263–278.

Zawadzki, M. and Jensen, T. (2020) Bullying and the Neoliberal University: A Co-Authored Autoethnography. *Management Learning* 51(4): 398–413.

9

AFFECTIVE, EMBODIED EXPERIENCES OF DOING FIELDWORK IN INDIA

A feminist's perspective

Nita Mishra

Introduction

This chapter draws on my long-term fieldwork experience as an ethnographic researcher working with women in rural communities in India. Re-visiting the field sites and exploring within my-self, using feminist methodologies, I show how researcher-participant relationships are continuously shaped and re-shaped. In such flexible and fluid contexts, there emerges a possibility of embodying affectual intensities between researcher-participants which researchers trained in Western scientific methodological rigours may miss or relegate to the backburner as I did in the writing of my doctoral thesis. Using my notes, photographs, conversations, narratives, poetry and memories in this chapter, I reflect upon what doing research really meant to me as researcher and to the women participants I met in different parts of Odisha during the fieldwork process.

The chapter is organised into three main sections. I begin with a summary of my doctoral research study. Next, I deliberate upon the methods used in collecting data, my experiences and conversations with women participants during my fieldwork, drawing on feminist methodologies to reflect upon the challenges faced and the learning taken from it. The final section concludes by reflecting on what this reveals about the lived experience of researcher-participant relationships and uses this to better understand what it means to do empowering research in specific contexts.

Linking my Irish doctoral degree with field study in India

My doctoral research[1] fieldwork focused on the lived experience of women 'duty-bearers' who have the responsibility to deliver food entitlements to poor communities under a rights-based programme, namely, the Integrated Child

DOI: 10.4324/9780429352492-9

Development Scheme[2] (ICDS) in Odisha State, India. A rights-based approach to development (Sengupta 2007) identifies a duty-bearer as someone who is obligated to a rights-holder who, in turn, has the right to demand fulfilment of a claim when it is denied or violated. This approach assumes that duty-bearers (including state officials and community workers) have obligations towards rights-holders (e.g., citizens and especially poor women in this study). A rights-holder can be an individual, a group of individuals or a legal personality. Scholars agree that the state (and its government) is the primary duty-bearer because it has the authority (1) to frame laws and regulations and (2) to enforce them with penal powers over all agents functioning within its jurisdiction and (3) also mediates between other states and coordinates with international institutions whose cooperation it may seek to fulfil human rights obligations (Sengupta 2007, pp. 328–329). However, studies on rights-based approaches to development have been limited to the role of non-state duty-bearers (Mishra and Lahiff 2018). The absence of in-depth investigation of state actors, as primary duty-bearers, poses limits to our understanding of how such approaches work in practice. More specifically, the challenges encountered, learning and lived experience of community-level duty-bearers have been largely overlooked by researchers and policy-makers.[3] My PhD research focuses on the lived experience of women duty-bearers at the community level (88 voluntary government employees and five civil society organisations) using qualitative in-depth interviews, narratives, group discussions, transect walks, photographs and observational data collected from 64 villages across four districts involving 180 participants.

My research interests lay in the qualitative study of "things in their natural settings, attempting to make sense of, or interpret, phenomena in terms of the meanings people bring to them" (Denzin and Lincoln 2005, p. 3). My approach to study was informed by ethnography and feminism which helped me to (1) establish linkages between the household and the state (Lister 2003), which is important because the state regulates access to resources that women need in order to change gendered relations of power at the household level; (2) be mindful (and thereof reflexive) in relation to institutional practices (Lazar 2005) that have implications for change in the social practices of individuals; and (3) capture the experience, insights and perceptions of women duty-bearers through adopting an interpretive perspective. An interpretive perspective "explains how certain conditions came into existence and persist" (Stringer 2007, pp. 95–96).

Upon returning from the field to my desk in Dublin, I found myself surrounded by A4 notebooks with fieldnotes, brochures, pamphlets, organisational reports, pictures, recorded interviews and women's poetry about empowering themselves. I had extensive notes on details and descriptions of the people, places and the situated context of all participatory observation field sites, including what activities participants were engaged in during our meeting. Although photographs taken during the field study were not part of the thesis writing, they were an important support material for the final analysis.

Revisiting sites of fieldwork and the site of the self

Situating herself in multi-locations in her reflections, Lal (1999) paves the path forward for researchers "with crosscut, mixed and hybrid identities" (as a woman born in the global south living in the United States, a scholar and graduate student in the academy, a feminist and middle-class researcher in India) by questioning "prescriptive methodological guidelines, canonical texts, and authorizing forms of discourse" (p. 102–105). Using vignettes to illustrate my in-depth conversations with women participants, here, I reflect upon my own positionality during fieldwork in Odisha to make sense of not just the content of my interactions with participants (through interviews, group discussions, observations) but also of the nature of those interactions. Increasingly I was negotiating between insider-outsider positions – an insider to any Indian context for my Irish colleagues, and yet an outsider for the villagers in Odisha. In spite of the knowhow of language and belongingness to the region by birth, I was an outsider for the Odia villager in terms of my education and lifestyle, and this hybrid identity impacted the researcher-participant relationship in various ways during the fieldwork.

Scholars such as Bhattacharya argue that research guidelines must be adapted to include fluidity and messiness of qualitative research studies. Processes of consenting are easy, insider-outsider kinship relations between the researcher and the researched are fluid, and, when juxtaposed with Western structures of scientific inquiry, they create spaces of blurred relationships, messy methodology and collaborative designs (Bhattacharya 2007, p. 1095). To explore the meaning of these fluid relationships, characterised by continuous shaping and re-shaping, I re-visit the 'site' – the geographical site and the emotional 'site' of the fieldworker self – using feminist methodology to 'see' the fieldnotes, journal entries, photographs and anecdotes in addition to data collected through interviews, group discussions, transect walks and observations. I reflect upon "the combination of the *whats* and the *hows* that provide the cultural frame by which people make sense of their experiences", where the *whats* refer to the content and the *hows* refer to the interactional element in the interviews (McLachlan and Garcia 2015, p. 200). Additionally, I was aware that research encounters are always emotion-laden (different bodies interact in different ways) and also underlined by power relations (Militz et al. 2019, p. 1). For example, as a researcher I responded differently to my participants in different field sites in terms of trust and my feelings of being an insider/outsider. Similarly, participants were influenced by time constraints, trust and their expectations of the researcher.

My main methods of data collection for the doctoral thesis were open-ended interviews, group discussions, photographs, transect walks and participatory observations. An important reason for narrating women's lived experiences is to "trigger resonance in the reader" and to "have access to and experience research encounters differently" (Militz et al. 2019, p. 2). I decided to draw attention to such narrations which Militz et al. conceptualise as affectual methodology to understand affectual intensities or "a process of critical writing, reflection, and rewriting about moments of resonance between different bodies and objects in the field" (2019, p. 1). Thus, open-ended interviews focused on listening to participant's individual stories situated in their wider social and cultural context. I spent a few nights talking and listening to indigenous tribal communities who had commuted from remote parts of the district in preparation for a public hearing organised by a network of community-based organisations in Koraput town. From that listening, I identified two young tribal women (Bhanumati and Jamuna), whom I asked to participate in an open-ended but in-depth interview. I was interested in questions that feminist researchers, such as Lal (1999) ask of the researcher: how does she know what she knows, how will she know if participants are asserting their agency, and how and where could she [the researcher and/or participant] effect change, if possible? (p. 103). For example, in an attempt to hold government officials accountable to problems faced by villagers, I was interested in Bhanumati's narration of the strategy undertaken by their organisation:

> In Kenduput, Gulliguda and Pandari maatta villages, we took the water from the wells, and demonstrated to the block development officer[4] (BDO) on his desk how dirty the water was. We also had the media with us, and it got printed in local newspapers the next day. The BDO had to take immediate action, and got the wells cleaned.
>
> *(Bhanumati)*

In another instance, three Adivasi Ekta Sanghathan (AES) activists went with 20 women of Panasguda and Thetediguda village to the BDO demanding extra water supply during the summer months.

> The BDO told us to go to the water supply department, the latter sent us back to the BDO. After 3–4 such trips to and fro, we got a chance to check the office register only to find that according to official data, they had dug boring wells in our villages! So we had to contend with that misinformation, and prepare to fight a long battle.
>
> *(Jamuna)*

As an engaged feminist researcher, I was curious to see what these narratives entailed in terms of time, distance and effort. So, I travelled to the remote regions of the mountainous Eastern Ghats in India to visit the villages of Kenduput, Gulliguda and Pandari maatta. I realised that to achieve their objectives, Bhanumati and Jamuna had to commute, by foot, between remote hamlets which took two or three hours, and convince women to collectivise, make time to meet locally between hamlets and commit to commute to the block development office (BDO) in Koraput town in buses which traversed the region once a day. This meant at least three days of leave from daily chores (household, food gathering, farm work, tending to cattle, collecting firewood, wage work etc.) to be able to reach the BDO office to register a complaint. If the relevant officers were absent on the day, it meant planning for the next meeting which meant more time was consumed. Further conversations with the women one evening during the public hearing process revealed that the wells were not cleaned or dug respectively the same year that the complaints were made by the women. Thus, while probing deeper, it was revealed that collective efforts of women can be time-consuming and frustrating, and strengthening resolve to demand rights to basic needs from local officials is a continuous process. Ackerley and True (2010, pp. 100–104) write that being your own person is a matter of style, and eventually our politics, personality traits and personal constraints lead us to adopt a particular methodology. Following the process of participating, listening and travelling to these women, I could finally feel their stories. I could visualise them. I felt their struggle. Upon reflection, I realised that this feeling of their stories and struggles were moments of affectual intensities between myself (the researcher) and my participants.

Open-ended interviews with Upma,[5] a key informant, took place over a period of five field trips to Balianta block in Khorda district of Odisha state. Upma was from one of the villages in Balianta and had been working with an NGO which ran gender-based projects in the region for over two decades. On one occasion, Upma took me as a pillion rider on her scooter on non-existent mud roads and we stopped to talk to women we met along the way. Most of these interviews were thus unplanned. I soon realised that the women were easier to talk with if I didn't stop the flow of conversation with my note writing. In one instance, Upma and I were invited (for interviews) to the terrace of a woman panchayat[6] leader who wanted to use the interview time to tend her spices spread on the terrace floor to dry in the sun. Hunched on her knees, she patted and rounded mixed lentil pastes before placing them neatly to dry alongside red chillies on an old sari on the cemented floor. The dry chillies and the lentil cakes were part of the everyday diet of families and also sold in the local market. In between the interview, we shared recipes and stories of lentil cakes made by women (including my grandmother) in my village.

In another key instance, Supervisor X,[7] whom I pursued for weeks, never allowed me to interview her either at her desk or in group discussions. She works under the Department of Women and Child Welfare, and her chief responsibility is to manage Anganwadi workers' (AWW) activities at the village level. Each supervisor is in charge of more than 10 Anganwadi centres (AWC) which are regulated by Anganwadi workers who implement services under the Integrated Child and Development Services (ICDS). Supervisor X watched me carefully during my field work when I spoke with the other women in group discussions. On the last evening, coincidentally, we were in the same carriage on a five-hour train ride back to the city. This incident is significant because we encountered sexual harassment on the train, which bonded us. In our attempt to find safe spaces, we disembarked from the train in search of the police in an isolated station where the train was derailed. Eventually we boarded a different train. The incident changed our relationship and she became one of my key participants. Later she revealed that her suspicions of me were fed by investigative journalism which unearthed many corrupt practices such as accepting bribes by Anganwadi workers and their higher officials at the block level. Supervisor X had provided me with invaluable insights through her behaviour, her avoidance, her office activities and minimal responses to my queries.

Using vignettes where researchers situate themselves within the story, researchers evoke "sensual knowledge and experiences" (Militz et al. 2019, p. 2) in the reader, and an indication of the extent of affectual intensities that have occurred between the researcher and participants. Thus, in my reflexive encounters discussed earlier, long after the fieldwork was over, I could still smell the sun-soaked lentil cakes. The memories of the dry red chillies still make me sneeze. And I can still feel the over-powering sense of fear which both Supervisor X and I embodied and experienced on that evening train ride from Balugaon to Bhubaneswar as discussed in the following.

Group discussions were used in situations where multiple participants discussed their lived experiences. As a feminist researcher, I was conscious of power dynamics revealed through language, and silences, of the women participating in group discussions where some were eager to talk and others were silent. Feminist research situates the researcher subjectively in the field site through her hybrid identity (as a woman, researcher, Western-educated). This 'situatedness' enables the researcher to observe people lost in gaps, silences, margins and peripheries (Ackerly and True 2010, p. 22). In self-help group meetings conducted on verandahs, Anganwadi centres[8] and in open community spaces, I was able to learn about the lives of extraordinary women (participants, workers, helpers, panchayat representatives, health workers and self-help group, SHG, members) of Balianta block who were actively engaged in community activities. Sometimes it was difficult to demarcate between different roles women played in the community. For instance, Z was a panchayat ward member, member of two self-help groups and was a health worker locally. If I had 'followed the book' and visited the panchayat office to interview Z in her role as a ward member, I would have missed out on details of how she functioned in her capacity as an SHG secretary or what it was like to be a health worker. Militz et al note that vignettes, and our response(s) to them, can resonate with the reader producing embodied knowledge through textual encounters, facilitated by the researcher's writing where she (me, in this case) positions herself within the vignettes. This evocation of emotion is proof that "research encounters are always emotion-laden and imbued with power, reproducing and legitimating social hierarchy" (Militz et al. 2019, p. 1). In another village, female Sarpanch[9] Y recited poetry co-written by four other women which described empowerment and steps taken by them to empower themselves. The poem, read under a Tamarind tree, spoke of actions leading to banning female foeticides and encouraging daughters to enrol in schools and to be active in local elections, and was an attempt to spread awareness on gender-based violence. This poem was read out in their meetings to remind them of the challenges faced as women and girls in their community. I found that being able to find a listener as well as the process of reading it aloud to outsiders like me (literate, educated female researcher) was empowering for Y and her co-poets.

Photographs of meetings helped me record the sequence of events, places and identify participants who spent more time talking with me than others which were of immense help to fill in the gaps during data analysis. In her work on visual ethnography, Pink (2001, p. 1) writes "just as an image might invoke a memory of an embodied affective experience, experiences also inspire images". Sifting through the photographs of a meeting at the Nakhara village Anganwadi centre, a photograph of a woman in a blue-bordered white sari staring squarely into the lens brought back the afternoon into life. I rummaged through my fieldnotes and diary and found my poetry. Before I go into the poetics of the encounter, I must tell you the story of Laxmipriya Ojha, the woman in the blue-bordered white sari before she became an Anganwadi helper. With tears in her eyes, Ojha described being abandoned by her husband, who left her for another woman as she could not conceive children. She went onto explain how she was then thrown out of the parental home by her brothers, who did not want to acknowledge her inheritance rights to the family property. Homeless for many years, Ojha survived on the streets as a rag picker, house-maid amongst other things. It is not uncommon for destitute women to undergo violence on the streets in the state. At a time like this, the government announcements for employment opportunities as Anganwadi workers and helpers with a focus on single abandoned women, as below, offered a lifeline to vulnerable women. The quotas for vulnerable women were ranked as – "widow (10%), unmarried women above 35 years of age (10%), orphan girl (5%), physically handicapped woman (5%), destitute or deserted woman (5%), Intermediate of Higher level of educated women (5%), and Scheduled Caste/ Scheduled Tribe (10%)" (GOO 2007).

This announcement combined with a very supportive local community helped Ojha[10] get the post as a helper for the village Anganwadi centre. I wanted to share the story of Ojha also because I want the reader to interact with the photograph (Figure 9.1) and its story to interpret it for themselves. "Meanings to texts are also given by its readers and not only what ethnographers ascribe to it" (Pink 2001, pp. 181–182). Without the photograph, Ojha's story would have been lost in the bigger picture of the thesis. What would also be lost is the importance of government policies which encourage women in vulnerable situations to apply for paid work.

> Tell my story. I live on the delta near the canal area. My *jhuggi* (hut) is under a plastic sheet tied to the sewage canal pipes on one side, and bamboo poles on the other with cloth sheets covering the entrance. This basti is my home now, and I am very grateful for having been chosen as the AWH by the basti people.
>
> *(Ojha)*

Photographs have a way of conveying messages to its readers. Every time I look at the photos, I found new information which I had overlooked. Photographs reminded me not only of the subtle power dynamics of the meetings but also people's vulnerabilities and anxieties. Photos of the women I talked with gave me faces to names I had noted down. Combined with observing and listening, photographs gave me clues to a woman's life, her thoughts and her priorities. These were

FIGURE 9.1 Laxmipriya Ojha, Anganwadi helper

not very different from mine. I was also reminded of how "the visual is inextricably interwoven with our personal identities, narratives, place, reality and truth" (Pink 2001, p. 1). My fieldnotes revealed much more. Things that were not possible to include in the thesis because the advice was to stick to the conventional way of writing were in my notes and in my poetry nonetheless.

    . . . and yet again
    in the hiding, then
    Stigma followed . . .
    . . . This time round
    I wrapped it around
    Stigma was I
    and I was Stigma
    . . . Stigma ceased to follow.
    (Mishra 2018, pp. 569–570)

Transect walks were a key method in my field research. I adapted the use of 'transect walks'[11] across villages as a preparatory step to understand who lived where and why, to understand the caste and gender dynamics of community life and to immerse in the everyday village life. Transect walks have been used in research projects especially in designing urban neighbourhood development projects by mapping localities to understand what can be done in which geographical area. For example, Powell discusses an interdisciplinary project where students walked around in El Chorrillo neighbourhood in Panama City to "account for the ways in which people construct meaning of space and place through personal experience and everyday activity – the lived experience, in other words, of place" (2010, p. 542). As such, the transect walking method went beyond mere documentation and mapping of streets, buildings, roads and other structures in the environment by uncovering narratives and lived experiences of the El Chorrillo community. In a study of public health, medical undergraduate students were introduced to transect walk methodology to encourage participatory education through a bird's eye view of a specific locality and facilitated problem identification and how existing local resources could be mobilised to solve problems (Dongre et al. 2009).

While transect walks originally refer to predetermined routes,[12] I adapted the method to explore different parts of villages where I had not been earlier. For example, in Macchkund block, through my walks into unknown (to me earlier) territories, I realised that the *Prathmik* schools and the Anganwadi Centres were three kilometres away from the nearest village with little or no public transport. During my transect walk in Martalaba village, I came across a small white church newly built atop the village hill. Later I was able to examine critically the role of the church through open-ended interviews with the villagers, which revealed that an important strategy advocated by local civil society organisations was to re-claim tribal heritage as opposed to those being imposed by 'newer' religions. Transect walks also revealed that women and men used the local ponds for all their water-related needs at different hours of the day. Observing and talking with local villagers in Odisha gave a glimpse of how people lived in certain areas according to their caste and religion. In Banapur block, multiple transect walks gave a clearer view of how different ethnic and caste groups have built their own hamlets within the villages. For instance, the Madrasa Islamia school was located within the Muslim sahi, and only Muslim children were enrolled in this primary school. Interestingly the Muslim sahi is located in the heart of the Hindu town of Banapur and opposite the 13th-century Hindu temple of Dakshya Prajapati. Another neighbourhood on the outskirts of the village was that of the *Panikhiya* caste, a designated backward caste but people of this caste were allowed into Brahmin homes to help in household chores and the kitchen. As Dongre et al. cite,

> Walking together . . . with local people removed barriers in communications and villagers expressed their views on different aspects of the public health topic more candidly, which does not usually happen in a community meeting or even in a one-to-one interview.
>
> *(2009, p. 1093)*

In my study, transect walks provided an opportunity to situate myself, observe and participate in ongoing village events. It also enabled a more participatory approach whereby curious onlookers became interested in my study and willing participants.

In Cuttack city, my transect walk in Jagannathpur *basti* (urban slum) led me to the (erstwhile) Leper's colony (Robertson 2009) at Gandhipalli built in 1919 and supported by the British raj. Within high compound walls, there was a self-sufficient community, urban slums, a parish church (the Baptist Missionary Society) and a primary school. I walked towards sounds of children chanting the English alphabet into a room where a young woman sat on a chair with 12 children (5–8 years of age) on the floor. The teacher asked me to be seated on a wooden bed across the tiny room. No documents on ethics were required here in my unplanned visit to the classroom. The encounter was guided by mutual awe and respect for the other. I had open-ended interviews with Lipina, the teacher.

An important tool of data collection was the use of participatory observations in all field sites because it enabled the research to gain access and in-depth information on how duty-bearers secured rights to people in their own settings and words. Gold (1958, cited in Matthews and Ross 2010) has drawn a scale of participant observation, ranging from the complete participant, the participant as observer, the observer as participant and the complete observer. A complete participant becomes a complete member of the group being researched, and a complete observer is totally detached from the group being research. The participant as observer "reveals her presence and her research role to the group", and in the observer as participant mode, "the researcher is moving away from the idea of participating" and data collection becomes more formal and structured (Matthews and Ross 2010, p. 258). In my field study, at any given point of time, I was on this continuum more as 'participant-observer' and 'observer-participant', which I have referred to while analysing data. I was a 'participant-observer' was at the public event of the One Billion Rising[13] (a global campaign to end gender-based violence) in Tangi village. As a speaker, I shared my views and experiences of being a girl child and a woman constantly searching for safe spaces. As a researcher, I jotted down notes on what the women spoke of, the school girls' speech, and observed how the choice of the market space to have the event attracted the men to pause and listen. Participating in the event provided me first-hand experience of and insights into the best possible way to break the silence around gender-based violence in the local community. This was a rare opportunity to observe, listen and share experiences and information.

In the late evening meetings with tribal women who had gathered for three days at the AES office premises from different parts of Koraput district to participate in the public hearing process, I was a 'complete-participant' as I engaged in preparing the food, baby-sitting, cleaning and engaging with other daily chores in the NGO office which housed almost 50 people. All the while I was listening. I have thus used the context to refine my tools of data collection, especially when interacting with people at the village level. Bryman (2008) writes that it is sometimes difficult to distinguish between ethnography and participant observation as both entail immersion in the research site and the social life of the research subject. In the case of ethnography, it would require an extended involvement, and Bryman argues that for many scholars pure ethnography would mean no reference to other documents or conduct of formal interviews. In this, however, my research differs, and participant observation is combined with interviews – both formal and informal as well as the collection of documents, while the immersion in the field was for a continuous long period of time. In more than one way, my positionality is akin to Lal's, where she notes how difficult it is "to retain the boundaries . . . of academic inquiry" (Lal 1999, pp. 104–105). As Lal leaves behind her initial research design in favour of ethnographic observations and open-ended interviews (p. 105), my field sites were increasingly guiding my methodologies emphasising fluidity, messiness and affectual intensities.

Opening up trust and lines of communication

I visited the sites more than once. In the case of Banapur, I made multiple visits because of the uniqueness of the case for the study. The Banapur case also illustrates how trust and the lines of communication open up when someone knew somebody who knew a cousin in another village to establish my ancestral links to the region, and therefore credibility to what I had to say or do. Things moved faster after that. My credentials were linked to the digging of my roots, not to the paper certifying me as a PhD student in CUBS.[14] The latter had no standing.

Having figured out my ancestry (land owners, Brahmin caste and priestly ancestors), women workers would offer me chairs to sit for informal interviews while I insisted (and succeeded) on sitting on the floor wherever they were. Caste divisions could have cast a shadow on my interactions with women if I were from a 'lower' (lower in social hierarchy) caste. But being from a family of an accepted higher (in social hierarchy) caste and requesting to talk with women from other castes made the latter curious about the reasons behind the request. For many women it indicated humility in my actions which drew them to me. Trust was also generated by simple gestures such as waiting for hours until women finished their work including household chores to have group discussions. Sometimes Anganwadi helpers came to me at night to avoid being seen by others because they felt better to share information without being seen to do so by the villagers. At other times, I would walk with them to and from the Anganwadi centre, listening to their tales of increasing burden of work at home and at the centre.

My greatest asset was my knowledge and use of the local Odia language. I had no interpreters in Banapur unlike in a few villages in Koraput district where numerous tribal dialects were dominant, and I needed an interpreter to a great extent. My dependence on an interpreter to make sense of what participants were saying sometimes led to confusion and I needed to validate information received. In a way I felt I had to develop trust with my interpreter, and in such fieldwork situations, I felt like an outsider. The ability to speak and understand Odia helped me to have a direct relationship with my participants. I could understand, discuss and have long conversations with participants to get my perspective on any given situation right. Women were always interested in knowing more about me and were able to form individual relationships with me. I was also never short of time for them. I genuinely wanted to know their story and share mine. Feminist methodologies privileging participation, representation, interpretation and reflexivity (Byrne and Lentin 2000) also help deconstruct power relationships between the researcher and the researched (Varga-Dobai 2012) and facilitates an underlying political commitment to the emancipation of women. In two different instances, I realised women (duty-bearers and rights-holders) were seeking agency through asserting themselves. In such scenarios, the participants held a powerful position which the researcher could only feel, observe and narrate. In my case, for example, one instance was when the unionised anganawadi workers and helpers held multiple rallies and demonstrations outside the BDO office in Banapur and in the capital city of Bhubaneswar. I walked with them. I interviewed many workers. I have video footage of the events. The second instance where I witnessed women asserting themselves was through my in-depth open-ended interviews with women beneficiaries of the ICDS entitlements where their agency was reflected in their understanding of their rights, their assertion through collective protests and demands for the suspension of corrupt AWWs and child development project officers (CDPO).

On one occasion, I was out in the paddy fields talking with women while they were busy harvesting the crop. I had my meal with them – rice soaked in its cooked water (rice-beer) and onions. We chatted. They asked me more questions than I did on that particular day. And then popped the request: "*Will you please stand for state elections from our village? We will make sure you win it*". Having the ability to know "when should my researcher self kick in [or not]", as Bhattacharya ponders in her own study, was crucial in this context (2007, p. 1096). This was a very conflicting moment for me as a researcher and as a woman whose ancestors were cremated in the region. Participants' expectations of me were not surprising but the level of trust in my capabilities overwhelmed me. Elsewhere, Bhattacharya writes, "I want to shout out loud, No, don't trust me, please". Granting me such privilege becomes more of a burden than a relief (Bhattacharya 2007, p. 1098). I was honest in my response. Thanking them, I said I could support their agency in other ways such as share information on entitlements and discuss bureaucratic processes if they needed. I narrated the story of my widowed great grandmother who walked across three districts of rough terrain in her early twenties with a son and two grandsons to reach these villages to start a new life. The next time we met, the women said:

If your grandmother could walk all those miles in the 1940s, and if a girl from our villages (*amma gaon ra jhiya*) can go as far as Europe why cannot we stand up for our rights in our own village.

(W)

Sharing stories were empowering moments for both the researcher and participants as both inspired and encouraged each other. In the earlier examples, one observes shifting dynamics of power between researcher and participants where the former is a mere observer, documenter and/or facilitator of an unfolding process of empowerment.

One of my most memorable interviews was a video interview of a young local councillor in Banapur. The councillor belonged to the fishing community (caste) and won local elections using the 33% reservation quota offered to women of vulnerable communities by the Indian Constitution. If a researcher went by numbers of women in the local municipality office,[15] it would be easy to say that women were empowered. However, during the in-depth interview process, the councillor broke down in tears and revealed that her father used violence and forced her to stand for elections to realise his own dreams of an equal society. I held the young woman and decided to erase the video footage of the interview. Was I, like Bhattacharya pondered,

> willing to sacrifice academic rigor for maintaining the blurred relationships of sisterhood, friendship, mentorship, and the researcher and researched? How else can academic rigor and trustworthiness be redefined when consenting, kinship relations and shared cultural understandings intersect in transnational feminist research?
>
> *(Bhattacharya 2007, p. 1098)*

This in-depth interview with the young councillor reflects on the affectual intensities between the researcher and her participant, and how researchers let the site determine the course of the study. It also shows that empowerment processes are complex. Although winning local elections and attending to affairs of the municipality can be exemplified as women's political empowerment, a closer look reveals that the accepted norm was for the menfolk (fathers, husbands and brothers) to accompany the elected women to municipality meeting. Was that empowerment? Perhaps 'yes' because in spite of patriarchal norms, new spaces have been created through legislation which allow and encourage women to step out of homes.

Lal (1999) argues that during (ethnographically) detailed, open-ended interviews, the dichotomy between the researcher and research subjects becomes merged, and interviews take their own shape and logic. In cases where in-depth conversation occurs, the time period in terms of the number of days of conversations and observations is less important ethnographically than the information and insight gained from them. I decided to let the sites determine my methodology as "methods in the hands of feminist researchers begin to take on a new context of practice" (Hesse-Biber 2010, p. 171).

Once I walked all afternoon and reached an Anganwadi centre which had just closed down after the days' work. The worker and the helper were sitting by the pond opposite the centre, chatting with other women in the village. It was 3 pm. Women were free from household chores and it was not unusual for them to sit out by ponds to catch up with local news. That is how news about this '*amma gaon ra jhiya*'-type researcher spread to other Anganwadi centres and opened doors for my interviews with large numbers of workers and helpers. I sat on the steps leading to the pond, and we started a conversation which led to my showing photographs of my family to women who were complete strangers. They were also curious about meeting me. Shyly, a few women asked me personal questions:

> Are you a journalist, a government official distributing post-cyclone relief material, or enlisting names for more entitlements?
> Are you married? If yes, then why don't you have the vermillion on your forehead? Are you a Hindu as your second name suggested? Do you have children? Who looked after them while you are away for weeks altogether? Are you serious that your partner cooked, cleaned and looked after the children in your absence? He also goes to an office?

The questions gave me a perspective on who they thought I was. I had no hesitation answering. We exchanged phone numbers. Trust followed gradually. I found this as an effective way of building relationships with my participants in contrast to making a list of Anganwadi centres and conducting formal interviews in all 80 centres.

Researcher-participant relationships

The researcher-participant relationship was of primary concern to me. In most cases, the researcher-researched relationship has been long term and fluid. Many times, I found myself responding and adhering to participants' needs, requests and contexts, thus blurring the lines between researcher-participant relationship. In my participatory observations, I informed participants that I was interested in participating in whatever they were doing so that I would not waste their time, and yet I also chatted informally with them. This led to long walks across the villages with the duty-bearers shadowing them in their daily activities, their meetings and in protest marches. The researcher-participant relationship went through difficult phases where the women of two villages expressed their wish for me to stand for political elections to usher in social change. Raising their expectations was not the aim of the research, and yet it was an opportune moment, as a feminist researcher, to discuss various options of political empowerment for the village women. Raising expectations of participants which may lead to disappointments problematises good ethical research in Western universities as researchers can be charged with deception, which is morally unacceptable (Christians 2005, p. 145). Keeping in mind the possibility of unforeseen deceptions, and any other harm which my study may bring, I constantly brainstormed with my participants about various possibilities which may bring in desired change for them.

> The important thing here, however, is not to promise what you cannot deliver. There should be some kind of 'contract' between researcher and informants – not necessarily a written document, but a series of implicit and/or explicit agreements by which the researcher is bound.
>
> *(Le Voi 2002, p. 161)*

Heeding Mies's (1983) call for a conscious partiality leveraging subjectivism and empathy as opposed to value-free research and indifference towards research participants, my aim was to create emancipatory value during my fieldwork where possible. I also felt morally responsible to support women's agency. Therefore, while I declined the offer to represent the women in local elections, I found it useful to facilitate discussions on the laws and policies related to entitlements to food, housing and health care of my participants. It was a starting point for discussions on the 73rd and 74th constitutional amendments, wherein reservation of seats for women in local political bodies was made legal. Laws et al. (2013) have argued that facilitation was a good way to manage great expectations of participants. Microanalytic readings of everyday practices (Phoenix and Pattynama 2006, p. 190) opened spaces for political dialogue. Furthermore, boundaries of dialogues should be determined by common political emancipatory goals, and the tactical and strategic priorities should be led by those whose needs are judged by the participants of the dialogue to be the most urgent (Yuval-Davis 2006, p. 206).

I found that researcher-participant 'power' relationships were fluid. Factors such as individual personalities of participants, official position in the government department and international exposure to development issues made a difference to the way officials approached an issue and the interviews. In some instances, geographical context and personal life histories determine the relations between the researcher and participant. As said, language helps merge the boundaries between the researcher and the participant. Acknowledging "the shifting or fluid rather than fixed divide of insider/outsider status", Gair (2012) argues that "a simplistic dualism of researcher as either an empathic insider or an ill-informed outsider" can ignore the common quest to hear, feel, understand, and value the stories of others" (2012, pp. 138–140). Sometimes researchers like me may find themselves in very different roles right in the middle of a study:

> as an insider (in Banapur), and as an outsider (in Koraput), and a bit of both in Balianta; situations which were determined by knowledge and use of language, customs, clothes, accent, and ancestry in my case; my own worldview which interacted in every field site on how I saw myself.[16]

This led me to question the effects of my presence in the field at all times. In the villages of Koraput, I was more of an outside researcher in the conventional sense of the term. In Balianta villages, I developed close emotional bonds with some participants in spite of my inability to help with job opportunities as requested by a couple of women workers. Interestingly, while I strongly felt as an outsider because of my Western university education, Banapur villagers accepted me as an insider as '*ama gaon ra jhiya*' because of my ancestral links to one of the villages. Added to those kinship links was an active relationship with the villages through summer visits, festivals and other events, which made my status as an insider valid to a great extent.

Shifting power dynamics between the researcher and the participant provide insight into the intersectionality of both social actors in the research process. For instance, if I had stopped short of pursuing Supervisor X because of her evident reluctance in being openly interviewed, in tandem with strict rules of consent and other 'ethical concerns' outlined in Western universities instead of following my instinct for interviews, I feel my study would have been incomplete. Scholars like Bishop (2005) direct attention to the need to challenge modernist discourses with their features of objectivity/subjectivity, replicability and external measures for validity, which is difficult for indigenous researchers (such as my hybrid identity) because these measures of validity are positioned/defined within another (dominant Western) worldview (p. 127). More recently, Bell and Kothiyal critique "research ethics [which] encourage researchers to abandon local, contextually relevant ethical principles in favour of Western ethics codes" because they may not be universally beneficial and may be interpreted differently in different cultural contexts (2018, p. 549). Following one's instinct in fieldwork, therefore, meant taking into cognizance the situatedness, geographical location, (dis)trust and other specific incidents which (may) shape responses and relationships between the researcher and participants.

The difficulty of obtaining written consent from participants because they were wary of signing any documents may be regarded as a failure to adhere to strict ethical norms by conventional research methodological norms. After trying to do so, in a few places, I decided against it, also because it would ward people off me. It is also not a common practice in India to do so. Participants were more at ease with introducing me to each other as '*ama gaon ra jhiye*' (daughter of our village) and trusted me with their stories. I informed participants at all stages and explained what to expect. During group discussions with women's self-help groups, I described my intentions, and where I came from, and asked to be questioned by them. Most participants were very interested in knowing about my personal history, and that added spice to the interviews. In many instances as in the three-day preparation of the public hearing event in Koraput, the meetings between Upma and the women in Balianta, and in the Anganwadi workers' political demonstrations outside the Banapur BDO office and the Odisha Secretariat in Bhubaneswar, as a researcher, I was a mute spectator to the unfolding of historical events. I was participant, observer and a witness to moments of 'agency' and empowerment of the researched.

Conclusion

This chapter highlights a personal journey of learning and reflecting during the research process. Using a combination of anecdotes of lived experience of the researcher-participant relationship, and scholarly works on feminist methodologies, visual ethnography, poetic inquiry and affectual methodologies, I have situated my own experiences of field research and reflected upon my interactions and relationships with participants. I learnt that my story was not very different from theirs. I learnt that empowerment is a journey linking the personal with the public. I learnt that concerns of safety in public transport at night were as daunting to my participants as it was to me! At such moments, reluctant-participants can become co-dependent in search of safety, and I learnt valuable aspects of participants' silences which would not have been possible in the structured research scenario. I learnt that many a time my training as an academic researcher did not equip me with the wisdom to deal between conflicting emotions of empathy and neutrality in an interview context. At such moments, my spirituality came to the rescue, and I followed my humane spirit. I ended holding hands of a heart-broken woman switching off the recorder or handing cash to a dejected poor father of a newborn baby who had no money to get his wife discharged from a government hospital. I learnt that violation of people's rights to free health care cannot be corrected by my research, and I have to act. I learnt that poor women sometimes break rules to re-define their duty-bearing roles, and that this was okay. I also learnt that so long as it helped another poor woman or a whole community, it was okay, and that it was not my mandate to report such breaks. An important lesson learnt was that if community-level women duty-bearers decided to follow a certain path under the given circumstances, it must be the most appropriate one. I also learnt that education or literacy was not key to empowerment but understanding ones' context and acting in accordance with that was crucial to 'agency-creation' for women. More importantly, I found myself a transformed woman valuing not the research process or the doctorate but the stories that I had heard and the difference I had made in the lives of a few women (and men) I met on the journey.

I was, time and again, advised to publish in Scopus-ranked journals to ensure high impact of my research. Interestingly, all such journals are based in the Western world and the waiting time for publishing seems unendingly long. I was in a hurry to publish because I felt my data would become redundant otherwise. Because I wanted my research participants to see their names in spaces which were of significance to them. Because for me sharing my research findings with my peers in India made more sense than in the journals published in the West. I wanted peer-reviewed acceptance of my research findings and discussions in the context I did my study. Does that make my study of less value? Perhaps 'yes' if I were to prioritise university appointments in the Western world. But Scopus rankings are a marked feature of exclusion, not inclusion, in academic work. So I decided that the Western world could wait.

Notes

1 My PhD awarded in June 2018 was titled 'Operationalising Rights-Based Approaches to Development: A study of state and non-state duty-bearers in Odisha'.
2 Services under the ICDS are mandatory under the National Food Security Act 2013, India
3 For more details on this aspect, see Mishra (2021).
4 Block Development Officers are state government employees in charge of development projects in their blocks which are administrative units below district level administration and above municipalities and grama panchayats in hierarchy. There are 314 blocks in Odisha. For more information, please see https://odishapanchayat.gov.in/english/demographic.asp. (accessed on 24/08/2020)
5 I have chosen, with due permissions, to name participants who wish to be identified because it makes a difference to their 'agency'. In other cases I have kept them anonymous as X, Y, and so on.

6 Panchayat means 'assembly of the five' which refers to grassroots democracy in India whereby the village is the lowest unit of government administration. See <https://pria. org/panchayathub/panchayat_text_view.php> accessed 23/08/2020. For more on panchayati raj, see Lakshmi (2016) and Chak (2017).

7 In the hierarchy of positions, Anganwadi workers' immediate superiors are called Supervisors who in turn report to Child Development Project Officers (CDPO)

8 Anganwadis' (courtyard) are community centres where the ICDS services can be accessed

9 Sarpanch is the head of a panchayat which is a group of 10–125 villages in a block.

10 Laxmipriya Ojha, Anganwadi Helper, Nakhara Anganwadi Centre, Balianta Block, Khorda district, Odisha, India

11 https://siteresources.worldbank.org/EXTTOPPSISOU/Resources/1424002-1185304794278/4026035-1185375653056/4028835-1185375678936/1_Transect_walk.pdf

12 https://sswm.info/humanitarian-crises/urban-settings/planning-process-tools/exploring-tools/transect-walk [accessed 25/08/2020]

13 www.onebillionrising.org/about/campaign/one-billion-rising/

14 Cork University Business School

15 Fieldnotes during PhD research

16 Fieldnotes during PhD research

References

Ackerly, B.A. and True, J. (2010) *Doing Feminist Research in Political and Social Science*. New York: Palgrave Macmillan.

Bell, E. and Kothiyal, N. (2018) Ethics Creep from the Core to the Periphery. In *The SAGE Handbook of Qualitative Business and Management Research Methods: History and Traditions*, 546–561. London: Sage.

Bhattacharya, K. (2007) Consenting to the Consent Form What Are the Fixed and Fluid Understandings Between the Researcher and the Researched? *Qualitative Inquiry* 13(8): 1095–1115.

Bishop, R. (2005) Freeing Ourselves from Neocolonial Domination in Research: A Kaupapa Maori Approach to Creating Knowledge. In *The Sage Handbook of Qualitative Research*, 3rd Edition, edited by N.K. Denzin and Y.S. Lincoln, 109–138. London: Sage.

Bryman, A. (2008) *Social Research Methods*, 3rd Edition. Oxford: Oxford University Press.

Byrne, A. and Lentin, R. (2000) Introduction: Feminist Research Methodologies in the Social Sciences. In *Researching Women: Feminist Research Methodologies in the Social Science in Ireland*, edited by Anne Byrne and Ronit Lentin. Dublin: Institute of Public Administration.

Chak, H. (2017) The Three-Tire Panchayati Raj System in India. *The Researchers' International Research Journal, Jharkhand* 3(1): 51–61.

Christians, C.G. (2005) Ethics and Politics in Qualitative Research. In *The Sage Handbook of Qualitative Research*, 3rd Edition, edited by Norman K. Denzin and Yvonne S. Lincoln, 139–164. London: Sage.

Denzin, Norman K. and Lincoln, Y. S. (2005) Introduction. In *The Sage Handbook of Qualitative Research*, 3rd Edition, edited by Norman K. Denzin and Yvonne S. Lincoln, 1–32. London: Sage.

Dongre, A.R., Deshmukh, P. and Garg, B.S. (2009) Using the 'Transect Walk' as a Public Health Teaching and Learning Tool. *Medical Education* 43(11): 1093–1094.

Gair, S. (2012) Feeling Their Stories: Contemplating Empathy, Insider/Outsider Positionings, and Enriching Qualitative Research. *Qualitative Health Research* 22(1): 134–143.

GOO (2007) *Revised Guidelines for Selection of Anganwadi Workers (Women and Child Development Department, Mona Sharma)*, Government of Odisha, India.

Hesse-Biber, S. (2010) Feminist Approaches to Mixed Methods Research. In *SAGE Handbook of Mixed Methods in Social and Behavioural Research*, edited by A. Tashakkori and Charles Teddlie, 169–192. London: Sage.

Lakshmi, C.P. (2016) Recent Scenario of Panchayati Raj in India. *Proceedings of the Indian History Congress* 77: 1016–1022.

Lal, J. (1999) Situating Locations: The Politics of Self, Identity, and 'Other' in Living and Writing the Text. In *Feminist Approaches to Theory and Methodology*, edited by Sharlene Hesse-Biber, Christina Gilmartin, and Robin Lydenberg, 100–137. Oxford: Oxford University Press.

Laws, S., Harper, C., Jones, N. and Marcus, R. (2013) *Research for Development: A Practical Guide*. London: Sage.

Lazar, S. (2005) Citizens Despite the State: Everyday Corruption and Local Politics in El Alto, Bolivia. In *Corruption: Anthropological Perspectives*, edited by D. Haller and C. Shore, 212–228. London: Pluto Press.

Le Voi, M. (2002) Responsibilities, Rights and Ethics. In *Doing Post Graduate Research*, edited by S. Potter, 153–164. London: Sage.

Lister, R. (2003) *Citizenship: Feminist Perspectives*. London: Palgrave.

Matthews, B. and Ross, L. (2010) *Research Methods: A Practical Guide for the Social Sciences*. London: Longman.

McLachlan, C.J. and Garcia, R.J. (2015) Philosophy in Practice? Doctoral Struggles with Ontology and Subjectivity in Qualitative Interviewing. *Management Learning* 46(2): 195–210.

Mies, M. (1983) Towards a Methodology for Feminist Research. In *Theories of Women Studies*, edited by G. Bowles and R. D. Klein, 117–207. London: Routledge & Kegan Paul.

Militz, E., Faria, C. and Schurr, C. (2019) Affectual Intensities: Writing with Resonance as Feminist Methodology. *Area*: 1–8.

Mishra, N. (2018) And Stigma Followed Me Everywhere. *Hypatia* 33(3): 569–570.

Mishra, N. (2021) Operationalizing rights-based approaches to development: Chinks in the armour observed through a study of anganwadi workers in Odisha, India. In *Poverty and Human Rights: Multidisciplinary Perspectives*, edited by S. Egan and A. Chadwick, 171–187. London: Edward Elgar.

Mishra, N. and Lahiff, E. (2018) We Are the Locals: The Operationalisation of Rights-Based Approaches to Development by Non-governmental Organisations in Koraput District, Odisha. *European Journal of Development Research* 30(5): 809–822.

Phoenix, A. and Pattynama, P. (2006) Intersectionality. *European Journal of Women's Studies Copyright* 13(3): 187–192.

Pink, S. (2001) *Doing Visual Ethnography: Images, Media and Representation in Research*. London, Thousand Oaks, CA and New Delhi: Sage, 196 pp.

Powell, K. (2010) Making Sense of Place: Mapping as a Multisensory Research Method. *Qualitative Inquiry* 16(7): 539–555.

Robertson, J. (2009) The Leprosy Asylum in India: 1886–1947. *Journal of the History of Medicine and Allied Sciences* 64(4): 474–517.

Sengupta, A. (2007) Poverty Eradication and Human Rights. In *Freedom from Poverty as a Human Right: Who Owes What to the Very Poor?*, edited by Thomas Pogge, 323–344. Oxford: Oxford University Press.

Stringer, E.T. (2007) *Action Research*, 3rd Edition. London: Sage.

Varga-Dobai, K. (2012) The Relationship of Researcher and Participant in Qualitative Inquiry: From 'Self and Other' Binaries to the Poststructural Feminist Perspective of Subjectivity. *The Qualitative Report* 17: 1–17.

Yuval-Davis, N. (2006) Intersectionality and Feminist Politics. *European Journal of Women's Studies* 13(3): 193–209.

10

FROM DOING, TO WRITING, TO BEING, IN RESEARCH

Amanda Sinclair

Introduction

How might we make the doing, the writing and our being – as researchers – more hopeful and energising to the participants we study and to the audiences we are seeking to engage? In this chapter, I want to respond to these, in my view, critical questions. But to do so differently.

I have agonised over how to write this chapter. Rather than offer a detailed critique of conventional research or arguments as to why and how we might do and write research differently, I offer here a series of personal stories and reflections. My approach is no criticism of the importance and the value of those who have contributed to the first task. There are long and rich disciplinary traditions which have critiqued dominant, positivist and conventional scientific ways of conducting research. They include feminist, critical educational, sociological, philosophical, Indigenous and anti-colonial research (Freire 1972; Oakley 1972, 2000; Harding 1987; Smith 1999; Moreton Robinson and Walter 2009). These bodies of work illuminate the damaging consequences of traditional research paradigms and customary methods, through which people are rendered objects (as well as subjects), cultural and gendered knowledge is appropriated and situated as inferior.[1] This kind of research often re-enacts colonising oppression and disempowerment while elevating the careers of the researcher (Pullen 2018; Phillips et al. 2014; Pullen et al. 2019).

This focus of this volume, in contrast, is on research that empowers. It calls us to interrogate how – or if – our customary processes of doing research empower those we research. But the question also applies to ourselves – as researchers – and what we permit for ourselves. Once I decided to write this chapter more personally, I felt empowered. It wasn't until I realised that my agonising over how to write it was symptomatic of precisely the phenomena that I wanted to disrupt that I found a

DOI: 10.4324/9780429352492-10

way in. I want to show – rather than write about or theorise – how research is inevitably lonely, fitful, full of self-doubt and set-backs, but also sometimes thrilling and uplifting. I want to show that it changes us. Indeed, I suggest that good research must change us. We've failed as researchers if we keep reproducing the same ways of studying, writing and being. As Sarah Gilmore and colleagues underline in the introduction to a Special Issue of *Management Learning* on Writing Differently, 'what and how we write is directly related with who we become' (2019: 9).

Writing from personal experience – albeit influenced by my reading of research and landmark texts – is also the only form of writing for which I now have the heart and appetite. The way this is written may be judged self-indulgent, overly confessional in tone. My work has certainly attracted those labels before. But researching and writing are, I believe, embodied experiences. I write it this way for you too, hopefully to enliven your memories, to free up your desires, your good instincts and your whole embodied self as a researcher, wherever you are at.

Doing research

My career as a researcher started in an ad hoc fashion, when I was coming to the end of my PhD. My marriage had broken down, and with two small children, I needed an income. I was invited to be part of a project exploring the experiences of women councillors in local government, including why local councils remained so trenchantly male-dominated.

The book describing the research, *Getting the Numbers: Women in Local Government* (1987), was published by a small Australian publishing house. Despite that it was original work and well-written,[2] despite that it was influential for more than a decade in encouraging women to stand for, and stay in, local government, it never seemed to count as 'real' research in the academic context. I frequently left it off my CV. I wasn't encouraged to think of this work as 'proper' research then, or maybe even till writing this.

For me as a young woman, research came with a capital 'R'. It was something that other people did, usually older men with big reputations, who commanded resources and teams of people. What I did was more modest, more partial, more personal. I didn't have a formal academic post. I depended on my mother and new partner to look after my children while I toured around the state and interviewed women and men about being on council. I transcribed and then put their experiences together in as compelling a way as I could in the book.

The process of this research and writing it up affected me deeply, as I have written about elsewhere (Sinclair 2019). It opened my eyes. It put me on a feminist path and fed a nascent, but tentative, activism. I wrote articles for newspapers. I wanted to get the message out into the public domain. I wanted to change things, and there was the dawning realisation that research could be, sometimes should be, political. Women in local government were, by definition, often isolated in country or suburban microcosms. They felt alone; they questioned themselves. Just hearing that they weren't alone was the first benefit of this work. The second benefit was helping them to understand that the exclusionary, occasionally abusive, experiences many of them faced weren't a failure of their own capacities or dedication. It was happening to other women who were – whatever their politics or ideology or backgrounds – disrupting the status quo simply by *being* women in council chambers. The women I'd spoken to and those coming after them deserved change, and I got a lot of feedback that our findings were empowering.

These experiences also irrevocably shaped my beliefs about who my writing should be for. It changed who I was as a researcher and began the – to use Laurel Richardson's wonderful phrase – 'de-disciplining' of my academic life (1997). I began to see conventional research, with its hierarchies, norms and prescriptions, as often alienating and disempowering for both the researched and the researcher. It wasn't just the excoriating rejections of journal manuscripts I later submitted – and there were many. Through experience – and beginning to read critical, feminist and post-colonial research – I began to see research could look different. It could be undertaken in ways that did not reproduce gendered hierarchies of research power.

I also understood with increasing clarity how the experiences of certain groups were ignored, left out or treated as not valid in the construction of knowledge. Carol Gilligan's carefully argued expose of models of moral development which privileged men's abstract rationality and universal principles was a powerful demonstration of how knowledge-building research excluded (Gilligan 1982; see also Harding 1987).

For the last ten or more years, I have had the opportunity to lead extended day-long workshops with groups of senior academic women at my university. Our focus is leadership, and as part of our work, these women share experiences as they develop their careers and identities as academics, teachers, researchers and research leaders. They come from a wide variety of disciplinary backgrounds, yet there are challenges which are common. I have also coached many senior academic researchers over the years. Some of them follow up after these workshops, some come from other universities and institutions, and they have often found themselves in research management roles.

Desperation and suffering are common companions for women navigating research careers. It's not the research itself that is punishing; rather, it is the pressures to demonstrate a particular, institutionally recognised and rewarded form of research and research 'leadership'. The conventional template of a research leader – especially in the sciences but also more broadly – is of a person completely dedicated to career, prepared and able to work huge hours, to forgo weeks of summer holidays to submit research funding applications, to travel often and for long periods and, to rest assured that, if they have families or other caring responsibilities, someone else will meet them.

The women academics I listen to and coach cite a variety of reasons why doing such research is close to impossible to enact, let alone debilitating and life-destroying to undertake. They literally have not enough time for it. Most get up in the dark to start preparing food or to care for younger and older generations. Many have heavy teaching schedules. They may have had an unconventional pathway; they may look small or speak quietly; they may attract biases and stereotypes about their youth or appearance that get in the way of appropriate recognition. When women try to identify these obstacles, for example with faculty superiors or in meetings, they find they are ostracised, labelled trouble-makers and sidelined.

Such constructs of research are damaging in the lives of academics (Bell and Sinclair 2014; Bell et al. 2019). They also potentially perpetuate colonising and neo-colonising knowledge practices and webs of resourcing (Connell 2014; Pullen et al. 2019). The 'best' research is that which attracts global capital and publication by American and British journals and editorial boards that, in turn, tend to reproduce elitist editorial practices (Metz et al. 2015). Research has historically, and is still, conducted largely in institutions in which power and capital are concentrated in the hands of a privileged elite and in which the different life experiences of women and culturally diverse participants and researchers are only intermittently registered (Criado Perez 2019).

In the face of such institutionalised structures and customs, are there ways to resist, steer around, come together to enact ways of doing, writing and being in research differently? From my coaching and teaching experiences, I've learned the value of several things. Firstly, it is to encourage others to 'de-discipline' their own notions of research. While all disciplines have their templates and measurement criteria, there is generally some scope for innovation, for example, by providing insight into why a research problem matters for oneself. Secondly, I encourage a sharing of experiences, which converts challenges from solitary sources of shame or humiliation to systemic or structural issues that can be named. A third strategy is to encourage researchers to model a more humane and multifaceted research leadership themselves. This may involve what I think of as 'identity work' (Sinclair 2016). For example, it might require asking ourselves the real risks of doing research differently. Most of us are familiar with – and can survive – journal rejections. We may have more choice than we think as to whether we collude with stultifying research norms. As I now go on to describe, despite the risks in challenging hierarchical and gendered understandings of the processes and purpose of research, we can celebrate and invite new research through living our lives, as researchers, differently.

Writing research

After the experience of *Getting the Numbers*, I took some teaching contracts and then began a part-time lecturing job at the University of Melbourne's Business School. I realised I needed some articles to progress in the academic world. I started small, local and with a mildly critical flavour, writing four articles published in Australian journals. These reported my early research on organisational cultures, management education and the importance of diversity. Like that first book, I found these articles didn't count either, in progressing above lecturer.

Next, I turned to my languishing PhD findings and sent a draft article off to *Human Relations*. I received a crushing rejection. I'd questioned the orthodoxy of the day – that teams were great organisational solutions without hierarchy, power or toxic emotions – and, clearly, I'd questioned the wisdom of the editor and reviewers too. I toned my critique down and was grateful to be accepted by *Organization Studies* – my first, and such a pivotal, taste of international recognition.

My early excursions into writing differently came in the form of including my own experiences in articles. These were driven by my emerging feminism, appreciating that some experiences are marginalised, but as researchers, we need to pay attention and record them. They were small rebellions of writing differently, and included 'Sex and the MBA'[3]. In this 1995 article I drew on my own experiences of teaching and working in a business school and being a mentor and confidante to women students. This article was published promptly in the new journal *Organization* which had an editorial board which included women and men who were sympathetic to the content and advocates for change in what were heavily male-dominated business school cultures.

I got braver and included more of my teaching experiences, including the move away from focusing on women and race in 'Managing Diversity' to the excoriating experience of introducing masculinities research to male audiences. I recently received a delightful email from a Norwegian researcher, Susann Gjerde, in which she quoted from a later article of mine on a related topic: 'Teaching Leadership Critically to MBAs: Experiences from Heaven and Hell'. The following excerpt begins with a quote from my 2007 article in italics, then adds Susann's:

"While I hope I have avoided the excesses of a confessional discourse, I believe it is through embodied, contextualized and critical accounts that new spaces for teaching leadership might be found." You are SO right, this is how new ways may be found, and your text was perfectly balanced in case you really wondered. I could feel your experience, learn from it, nod and laugh. It was very relatable and since you were so present in your paper it felt like a real conversation, and so I found myself wanting to respond.

Camaraderie on the journey of writing differently came with my visit to Cambridge in 2005 hosted by Chris Grey, and our many conversations wandering around college gardens, enjoying the jonquils and lamenting the dire state of management writing. We both enjoyed writing 'Writing Differently', which was published in *Organization*. Underneath or kicked along by the enjoyment of writing, this article had a serious intent. We wanted to challenge the abstruse and sometimes self-aggrandising writing that seemed to be prevalent, especially among critical management writers. Chris and I were talking the other day, and he took the view that despite some appreciative feedback from individual academics, our article hadn't shifted the norms much.

Since then, my own appetite for writing more conventional articles has sunk like a stone. Many of my own efforts to interrupt and interrogate management and leadership writing – to problematise how we construct a self through our writing; to experiment with fictional components; to disrupt gendered and pseudo-scientific or pompous writing – seem to have largely gone under the radar. For example, in 'Placing Self: How Might We Place Ourselves Differently in Studying Leadership?', I experimented with a fictional device – the two-part page – with the objective/conceptual argument on the 'top' half of the page, and the embodied, personal experience on the 'bottom'. I was seeking, I argued, 'to reveal the dualities and duplicities in discourses (of leadership), as well as the power and vulnerabilities that may be in play or repressed when we write about leadership'.

'A Material Dean', a fictional account of a woman Dean's experiences of leadership, and 'On breasts, knees and being fully human in leadership' seemed equally to mystify a small band of readers, rather than have the effect I was after, which was to encourage an injection of the physical into our accounts of leadership. No doubt the quality of the writing was inadequate to carry the experiment. Meantime, quite a few people were writing about writing organisational research differently (Pullen 2006; Phillips et al. 2014; Vachhani 2015). This, at least, was encouraging, and I felt a huge relief that other researchers were experimenting too.

I was influenced too, in my dissatisfaction, by my daughter's doctoral research looking at early modern Italian women writers and how they used their writing to construct selves – multiple selves – some compliant to classical authorities, simultaneously enabling subversive challenge to those authorities. Even getting read, as a woman author in the seventeenth century, was an achievement. The writing enabled a multi-layered construction of self that both established authority and defied attack. Reading about the skill and cleverness with which these early feminists adapted their writing – and the selves produced through the writing – I was struck by how predictable and uninventive we, in organisation studies, are about our writing.

How we write research matters – this is clear. But, of course, whether we feel able to write differently depends on power. As I acquired some small legitimacy as a researcher, my freedom to ignore conventions and push boundaries increased. I don't underestimate the constraints on more junior researchers, as Ruth Weatherall conveys so poignantly in her account of her drive to write differently her doctoral research on domestic violence (2019). In my own supervision of doctoral students, I've had to censor my urgings to write more radically or expressively, knowing that if students followed my advice, their successful completion might be jeopardised. Yet, as all students of poststructuralism understand, power and authority are never complete (Collinson 2020). There are always opportunities for resistance and agency in the gaps, lapses and disappointments left by conventions.

In a July 2016 interview in *The New Yorker*, philosopher Martha Nussbaum summarises conventional philosophical writing as 'scientific, abstract, hygienically pallid': its exhortations about how we should live rendered – paradoxically – cold and inert (Aviv 2016). Nussbaum recognises that all of us who write use writing to control life. We go on thinking, deliberating, phrasing, at least partially in refuge from fear. Nussbaum's antidote is to cultivate creative expression of emotion in our writing, thereby inviting readers to enter empathetically into other peoples' lives. I personally believe there is a huge hunger for surprising writing. I simply could not stop reading when I lighted upon the work of Audre Lorde, Laurel Richardson and Petra Munro Hendry. Insights exploded from the page, awakening my heart with resonances and the truths about human experience expressed.

Being a researcher, or perhaps being in research

In her beautiful conversational book with colleagues, educationalist Munro Hendry describes her own shift from understanding research as epistemological to ontological, from reconceiving research not as a knowledge product but as 'a process of being in relationship with others' (2018: 13). What does this ostensibly modest move involve? Munro Hendry says:

> to try and move more and more towards just being present. This notion of just listening and trusting that meaning will be made. That I don't have to analyse it, I don't have to turn it into data, and I don't do that kind of writing anymore . . . [by being] attentive to the fact that I want to honor our human experiences and what connects us as human beings . . . I'm not looking for anything anymore, I'm not striving to prove anything, to find out anything, I just want to be engaged in a process, and that's all I can do.
>
> *(2018: 141)*

Munro Hendry captures lucidly the final step that I explore in this chapter. It is from a focus on the *doing* of research to how we can *be* with others in our shared, but often different, quests for insight, accompanied by a dynamic mix of political, social and personal motives.

This move involves a shift in our roles and identities as researchers. It requires less attachment to research as instrumental or narrowly career-enhancing (Vachhani and Pullen 2019). It sometimes requires letting go of the need to impose – or even make – meaning from our research conversations. It demands we find ways to honour and respect the dignity and diversity of human experience, without re-rendering it in our own clever analyses or words. It demands we think again – and regularly – about who our research is for and how to genuinely empower those whose experiences we are recording and researching.

My own experiences – professional, personal and the whole terrain of life in between – have provoked and demanded I *be* in research differently. As I've described, and quite early on in my academic career, I remember asking myself, 'Can me listening, reading and writing about phenomena *qualify* as research?' Increasingly and through my own encounters with different forms of research, I realised it could! Here I share some recent experiences that have provided new and powerful demonstrations of how and why *being* different in research matters.

In November 2019, I was preparing to travel to India to speak at the invitation of a colleague, Lakshmi Priya Daniels, at the Stella Maris Women's College in Chennai (an all-women university), for the International Conference on Shaping Women's Leadership. Priya and I had met in 2018 when a colleague and associate dean, Professor Kate MacNeill, invited Priya to visit the University of Melbourne as part of a United Board Fellowship. Sharing an interest in women's education and leadership, the seeds for ongoing friendship and collaboration were sown.

As I often do, I was sharing my excitement about this visit with colleagues and former students. A friend of a colleague had done her yoga teacher training in Auroville, not far from Chennai. She inspired me to do some research. I found out a lot more about this UNESCO-recognised international community dedicated to operating by more communal, sustainable and spiritual principles, which was set up by The Mother and Sri Aurobindo in the late 1960s.

In a subsequent conversation with a favourite former MBA student, it turned out that one of her best friends lives with her partner and baby in Auroville. I made contact with this friend who provided a wealth of very useful 'on-the-ground' recommendations for accommodation, travel and yoga classes. She also offered to arrange a meeting with her bosses, the two founders of Eco Femme, a well-established Auroville social enterprise, which seeks to change social attitudes towards menstruation in India and equip women with re-usable (washable) sanitary pads.

On my second morning in Auroville, I got on my ebike and cruised the five minutes to the Eco Femme headquarters, a lovely old, shaded building with clusters of women sitting on the verandas packing the pads. I met with the founders, Kathy Walkling and Jessamijn Miedema, and Lauren Chockman, the good friend of my former student who works at Eco Femme, who brought along her young baby. We had no agenda, though I was hoping to learn more about their way of operating. They may reasonably have expected me to offer some value to them, given that I was a business school professor with an interest in women's leadership.

As I listened to them talk (and regretted not doing more prior online research of Eco Femme – they have an excellent website), the founders described a purpose and research-driven organisation. Their innovative, eco-conscious sourcing and their employee, educational and distribution practices had been designed to reach young Indian girls and women, especially in rural areas, where buying sanitary pads is expensive and contributes to the rubbish problem and where lack of knowledge and cultural norms meant girls felt a high degree of shame about menstruation.

Eco Femme's 'product' is much more than pads. Rather it is a sophisticated educational programme built on cultural knowledge and collaboration with local schools and community leaders. The founders have been generous with their product and design knowledge, and there are similar products on the market. No competitors have managed to emulate the scope of their distribution and educational networks. Guiding the organisation was a philosophy of empowerment: giving girls approaching puberty knowledge and autonomy to manage their own bodies, giving local women training to teach girls, engendering pride rather than shame about themselves and their bodies as women. According to one 14-year-old quoted on the website, her whole extended family now send the girls of the family to her as they approach puberty. She has become the family and community expert who mentors girls through this transition.

As we talked more, I asked the founders about leadership training. One mentioned that she had been sponsored to do some internationally accredited leadership training but had found it 'torture'. Worse than unhelpful, it had included psychological profiling. The templates that were offered in the programme of who was a leader simply weren't her.

The founders also volunteered that there had been interest in Eco Femme from academics seeking to undertake case studies. With the exception of one woman, these scholars didn't actually come and visit. One group emailed a copy of their published paper in a well-regarded journal but it was unreadable. All the richness and innovation in what Eco Femme has been doing and has achieved was crushed under the weighty application of institutional theory. Kathy admitted: 'I couldn't make much sense of it myself . . . [I]t was hard to really "find us" in that article'.

As a fellow academic, I was appalled. You've got to ask who is such academic writing serving? I had a copy of one of my books, *Women Leading*, which I had brought from Australia potentially to give, but I doubted its relevance. I certainly didn't want to impose an obligation to read it. These women had a lot of other important things to do. I scoured my brain for ideas that might assist Eco Femme do better than what they were doing – but could come up with little other than my admiration and support.

Back at the guest house, my Auroville and Eco Femme visit demanded another re-think of who and how I wanted to be as a researcher and leader. I had arrived thinking – I am ashamed to admit – that I was the expert researcher in women's leadership. The clear message to me from my visit was that in order to do empowering research – and especially in this very different cultural context – I needed to practice humility, to listen, learn and support.

These reflections, prompted by my Auroville visit, about my role and being – as a researcher – were given added momentum a week later at the Shaping Women's Leadership Conference in Chennai. Dr Priya Daniels, who had organised the conference at the Stella Maris College, wanted to expose the students of this university college to diverse inspiring speakers. We heard from several dynamic young women like Kirthi Jayakumar, founder of the Red Elephant Foundation, which assists homeless young Indians. As I listened, I heard not just activism and leadership but also systematic means of researching – empowering the life experiences of the people for whom the speakers were advocating, through cultivating deep understanding and mutual respect.

As another example, University of Melbourne academic Associate Professor Dolly Kikon described to the conference her innovative research methodologies exploring the violent consequences of militarisation on women and children in Nagaland (Kikon 2015). Dolly's research involved finding and visiting (sometimes at considerable risk to researchers and participants), listening to and recording heart-rending accounts of routinised, sustained violence experienced by these women, often at the hands of government troops. At the end of her presentation, Dolly shared with the audience a large box of her books. The research was written up not in a journal but as stories and accounts that invited the readership of those living the phenomena, those who might mobilise change from within cultures, and from the ground up. Traditional research and research writing erect false walls between research, women's leadership and political activism (Vachhani 2015). For me, Dolly provided a model of a very different – ethical and empowering – way of doing, writing and being in research.

Returning to Melbourne, many things intervened in my own processing of these experiences and writing this chapter. Several close family and friends died of cancer. The Covid virus struck the world, and, like many others, I had to learn to do my work and teaching online. I'd promised to send my Eco Femme friends a follow-up paper about my findings and the conference. Finally, I did, sending a draft version of the above reflection on my visit. I promptly received the following (edited) email from Kathy.

Dear Amanda,

It is lovely to hear from you and thank you for reaching out and sharing this work on your chapter with us for feedback and providing all the context. We are very happy to have you include the references to Eco Femme and I have gone through and made comments both to fine tune some accuracy and share some reflections on the piece which I found surprisingly moving to read. I actually found I had tears running down my face as I read it and on reflection realised there was something so affirming in what you wrote that really touched my heart.

And as I write these words, I realise I actually have no clue what I am going to say. . . . I just know I have to take a little risk. . . . I suppose it is because I also feel sure I can trust you because I am slowly starting to understand your inner orientation and what moves you in your work with women leaders and more to the point because of how you engaged and have continued to engage with us with genuine empathy.

I suppose the first thing that caught my attention in that little piece of writing was the focus on the shift from doing writing/research to being a presence through which the human experience can unfold and be received. There is something so organic and alive and real in that orientation and it quite literally thrills some part of my being to read things like this. It sounds like a 'modest move' that changes everything!

I have always found the words 'women's leadership' somehow mysterious – almost as though there is some conundrum in how these 2 words can even belong together. And I notice I am not the only woman perplexed by this. One thing that strikes me is how changeable women are. This is something I have contemplated a lot working on the theme of menstruation which is obviously about the embodiment of a cyclical experience – far from linear in nature! For years I battled with myself trying to fit in and be consistent which our modern world (yes even in Auroville) seems to demand – rigid 9–5 timings, consistent performance and relentless outputs. It is as though we are expected to perform as if we are in eternal spring. Navigating inconsistency and shifting inner dynamics seems to me to be one of the real challenges of being a woman and leading an organisation.

And this is really up for me these days as I am finding that I am simply unable to carry the mantle anymore of expectations that don't bring me alive – internal ones to be good and perform my duty as well as external plus I don't want to 'lead' anybody, never have – I prefer the notion of 'self-authorship'! . . . I suppose I am just musing out loud about why I feel deeply that old lenses of how we define leadership and relate these concepts to women and research and what are we actually doing as researchers feels so very interesting and relevant to me and I don't think there are easy answers as much as questions to be lived into. And your mail and chapter excerpt catalysed this stream of consciousness!

. . . I see that one thing you are doing is helping to re-write a story that enables real women and their life work to be validated and held respectfully, even if it is 'non-conforming' and I think that this has an immense value in our shifting, changing world where we need more compassionate and bigger spaces to hold our lives and our work within.

With love,
Kathy

Kathy also responded to the draft I sent of this chapter with the following observation:

It is often awkward to navigate professionals who assume to be experts and don't know what they don't know . . . mostly men as it happens! We have had a big learning curve about how to discern authentic openness which we are excited and happy to engage with and learn from and the kind of know-it-all types that think they are coming in to save us and help us, which is truly annoying to say the least. We tolerated years of it before wising up! It actually did help us, though, catalyze our own confidence in what we actually now know from lived experience.

Rather than being a leader whom I researched, Kathy became my teacher. In sharing her deeper and vulnerable feelings about 'being' a leader – especially as her own desires and appetites for the day-to-day work change – she has shown me new things. She has walked alongside, and nudged, me as I re-wrote this chapter. Could this be called research? Could our connection and conversations be called empowering methodologies? Certainly for me it has felt so – changing my sense of who I am and can be in research.

The above example is one of many where my teacher in doing and being in research differently has come from unexpected places – from research participants, students, friends, family. My sister-in-law, who recently passed away, was a drama educator. Over our long history together, but especially in the six months before she died, I read a lot of her work. She and her colleagues understand performance as potentially an empowering research method. Drama educators are especially interested in the potential of performance to express experiences, to move and change the understandings of audiences. Equally important are participants – students, community members or professional actors – who share authorship of the evolving story that is told (Bird and Sinclair 2019). Sinclair and Belliveau challenge researchers to bring:

the aesthetic and performative into their investigations of the social, cultural and political world; in so doing they highlight the potential to give voice to the marginalised, the silenced . . . and what's less visible in traditional academic research.

(2014: 5)

Final words

Why research? Why write an article or a book chapter to convey research findings? It's a question I ask myself a lot, especially lately, including in the writing of this chapter. A dark but perhaps accurate view is we research and write to fend off inconsequentiality: 'I come into being only in the context of establishing a narrative account of myself' (Riach et al. 2016). This is surely true of my own writing and this chapter.

But it is not all. A loving and compassionate view is that we research to share feelings and insights that may be appreciated freshly by readers; we write to change things, to help those who read our work feel courageous, hopeful and empowered (Richardson and St. Pierre 2005).

I've been looking through some old columns that I wrote as a regular contributor to an Australian business magazine. Many are actually pretty good, and all the better because, although they cover practical aspects of being managers and in organisations, they do so differently from the mainstream journalistic style. They don't exhort us to lift our game, to win or work harder. They are relatively free from – in fact critique – business jargon. They encourage openness and being with reality, including a meditation on facing death. After I wrote that particular column, my contract was terminated!

Yet reading those columns from over a decade and a half ago energised me. It reminded me of the importance of de-stabilising conventional notions of research and writing, of urging people to do and be in research differently. We must value research that doesn't show up in our journals but leaps out from the opportunities of the worlds around us.

Notes

1 I'm reminded of our recent bushfires in Australia and a whole tradition of expertise and research that ignored Australian Indigenous knowledge of fire and fire management.
2 Two co-authors, Margaret Bowman and Lynne Strahan, contributed a chapter respectively on history and women's organisations.
3 I have opted not to include full references for these articles and works as to do so would go against what I am arguing. My intention is, rather, to exemplify my journey in experimenting with writing.

References

Aviv, R. (2016) "'Captain of Her Soul' an Interview with Martha Nussbaum." *The New Yorker*, July 25.
Bell, E., S. Merilainen, S. Taylor, and J. Tienari. 2019. "Time's Up! Feminist Theory and Activism Meets Organization Studies." *Human Relations* 72(1): 4–22.

Bell, E., and A. Sinclair. 2014. "Reclaiming Eroticism in the Academy." *Organization* 21(2): 268–280.

Bird, J., and C. Sinclair. 2019. "Principles of Embodied Pedagogy: The Role of the Drama Educator in Transforming Student Understanding through a Collaborative and Embodied Aesthetic Practice." *Applied Theatre Research* 7(1): 21–36.

Collinson, D. 2020. "Only Connect: Exploring the Critical Dialectical Turn in Leadership Studies." *Organization Theory* 1: 1–22.

Connell, R. 2014. "Using Southern Theory: Decolonizing Social Thought in Theory, Research and Application." *Planning Theory* 13(2): 210–223.

Criado Perez, C. 2019. *Invisible Women: Data Bias in a World Designed for Men.* London: Chatto & Windus.

Freire, P. (trans M. Ramos). 1972. *Pedagogy of the Oppressed.* New York: Herder and Herder.

Gilligan, C. 1982. *In a Different Voice: Psychological Theory and Women's Development.* Cambridge, MA: Harvard University Press.

Gilmore, S., N. Harding, J. Helin, and A. Pullen. 2019. "Introduction to Special Issue 'Writing Differently'." *Management Learning* 50(1): 3–10.

Harding, S. 1987. *Feminism and Methodology: Social Science Issues.* Bloomington, IN: Indiana University Press.

Hendry, P.M., R.W. Mitchell, and P.W. Eaton. 2018. *Troubling Method: Narrative Research as Being.* New York: Peter Lang.

Kikon, D. 2015. *Life and Dignity: Women's Testimonies of Sexual Violence in Dimapur (Nagaland).* NESRC Monograph Series 1. Guwahati: North Eastern Social Research Centre.

Metz, I., A.-W. Harzing, and M. Zyphur. 2015, September. "Of Journal Editors and Editorial Boards: Who are the Trail-blazers in Increasing Editorial Board Gender Equality?" *British Journal of Management* 27(4): 712–726. https://doi.org/10.1111/1467-8551.12133

Moreton-Robinson, A., and M. Walter. 2009. "Indigenous Methodologies in Social Research." In *Social Research Methods,* edited by M. Walter, 1–18. South Melbourne: Oxford University Press.

Oakley, A. 1972. *Sex, Gender and Society.* London: Temple Smith.

Oakley, A. 2000. *Experiments in Knowing.* Cambridge: Polity Press.

Phillips, M., A. Pullen, and C. Rhodes. 2014. "Writing Organization as Gendered Practice: Interrupting the Libidinal Economy." *Organization Studies* 35(3): 313–333.

Pullen, A. 2006. "Gendering the Research Self: Social Practice and Corporeal Multiplicity in the Writing of Organizational Research." *Gender, Work and Organization* 13(3): 277–298.

Pullen, A. 2018. "Writing as Labiaplasty." *Organization* 25(1): 123–130.

Pullen, A., P. Lewis, B. Ozkazanc-Pan, and R. Connell. 2019. "New Maps of Struggle for Gender Justice: Rethinking Feminist Research on Organizations and Work." *Gender, Work and Organization* 26(1): 54–64.

Riach, K., N. Rumens, and M. Tyler. 2016. "Towards a Butlerian methodology: Undoing Organizational Performativity through Anti-narrative Research." *Human Relations* 69(11): 2069–2089.

Richardson, L. 1997. "Skirting a Pleated Text: De-disciplining an Academic Life." *Qualitative Inquiry* 3(3): 295–303.

Richardson, L., and E. St. Pierre. 2005. "Writing: A Method of Inquiry." In *The Sage Handbook of Qualitative Research* (3rd ed.), edited by N. Denzin and Y. Lincoln, 959–978. Thousand Oaks: Sage.

Sinclair, A. 2016. *Leading Mindfully.* Crow's Nest: Allen & Unwin.

Sinclair, A. 2019. "Five Movements in an Embodied Feminism: A Memoir." *Human Relations* 72(1): 144–158.

Sinclair, C. and G. Belliveau. 2014. "Editorial." *Journal of Artistic and Creative Education* 8(1) Special Issue on Performed Research: 5–7.

Smith, L. 1999. *Decolonising Methodologies*. London and Dunedin: Zed Books and University of Otago.

Vachhani, S. 2015. "Organizing Love: Thoughts on the Transformative and Activist Potential of Feminine Writing." *Gender, Work and Organization* 22(2): 148–162.

Vachhani, S., and A. Pullen. 2019. "Ethics, Politics and Feminist Organizing: Writing Feminist Infrapolitics and Affective Solidarity into Everyday Sexism." *Human Relations* 72(1): 23–47.

Weatherall, R. 2019. "Writing the Doctoral Thesis Differently." *Management Learning* 50(1): 100–113.

INDEX